Biochemistry and its Application

BIOCHEMISTRY AND ITS APPLICATION

Edited by:

Papita H Gourkhede

ARCLER
PRESS

www.arclerpress.com

Biochemistry and its Application

Papita H Gourkhede

Arcler Press
224 Shoreacres Road
Burlington, ON L7L 2H2
Canada
www.arclerpress.com
Email: orders@arclereducation.com

ISBN: 978-1-77469-505-0 (Hardcover)

ABOUT THE EDITOR

Dr.(Mrs). Papita H Gourkhede (1977) is presently serving as Assistant Professor, Department of Soil Science and Agriculture Chemistry, college of Agriculture, Vasantrao Naik Marathwada Krishi Vidyapeeth, Parbhani. She obtained her B. Sc.(Ag.) in 2001 from College of Agriculture , Nagpur from Dr. Panjabrao Deshmukh Krishi Vidyapeeth, Akola. Pursued M. Sc. (Ag.) in 2007 and Ph. D. in Soil Science And Agriculture Chemistry in 2012 from Vasantrao Naik Marathwada Krishi Vidhyapeeth, Parbhani. She started her career as Assistant professor in 2012. Her field of specialization is soil fertility, nutrient management, micronutrients, heavy metal remediation, Remote Sensing and Organic Farming. She has published 05 books and 35 research paper in National and international journals of reputed. She has participated in many National, State Seminar and symposiums. She has written 06 practical manuals for under graduate course for the benefit of students besides this delivered several Radio talk, Lectures in farmer traning programme. Dr. Papita has also written 112 popular articles in Agroone, Shetibhati, RCFsheti patrika and other reputed Magazines. She has received young Scientists award in 2016, ICAR national Award for research innovation in Dryland Agriculture in 2018, women Scientists Award in 2019, Agrocare Award in 2017, State level Best program officer award in 2020.

TABLE OF CONTENTS

LIST OF FIGURES

LIST OF ABBREVIATIONS

AATs	Alcohol Acyltransferases
ACF	Auto Correlation Function
ACTH	Adrenocorticotropic Hormone
AFP	Alpha -Fetoprotein
ALT	Alanine Transaminase
AP	Alkaline Phosphatase
APLAC	Asia-Pacific Laboratory Accreditation Co-Operation
ARF	Acute Renal Failure
AST	Aspartate Aminotransferase
ATP	Adenosine Triphosphate
ATP	Adenosine triphosphate
BRET	Bioluminescence Resonance Energy Transfer
BSI	British Standards Institution
CA	Cinnamic Acid
CFP	Cyan Fluorescent Protein
CHI	Community Health Index
CK	Creatine Kinase
CKD	Chronic Kidney Disease
CL	Cardiolipin
CPA	Certified Public Accountant
CSF	Cerebrospinal Fluid
DHB	Defense Health Board
DMAPP	Dimethylallyl Diphosphate
DNA	Deoxyribonucleic Acid
ECD	Electron Capture Dissociation
EDTA	Ethylenediamine Tetraacetic Acid
EF	*Enterococcus faecalis*
ELISA	Enzyme-Linked Immunosorbent Assay
EP	Enzyme-Product
ER	Endoplasmic Reticulum

ES	Enzyme-Substrate
ETD	Electron Transfer Dissociation
FAB	Fulfillment Assurance And Billing
FAD	Flavin Adenine Dinucleotide
FCS	Fluorescence Correlation Spectroscopy
FLS	Flavanol Synthase
FPLC	Fast Protein Liquid Chromatography
FRET	Fluorescence Resonance Energy Transfer
GC	Gas Chromatography
GFP	Green Fluorescent Protein
GFR	Glomerular Filtration Rate
GGT	G-Glutamyl Transferase
GLC	Gas-liquid chromatography
GST	Glutathione Transferase
HETP	Height Equivalent to A Theoretical Plate
HGT	Horizontal Gene Transfer
HPLC	High-Performance Liquid Chromatography
Hz	Hertz
ILAC	International Accreditation Co-operation
IMS	Information Management System
IPP	Isopentenyl Diphosphate
ISC	Intersystem Crossing
ISEs	Ion-Selective Electrodes
K	Kelvin
LC	Liquid Chromatography
LD	Lactate Dehydrogenase
LD	Linear Dichroism
MALDI	Matrix Assisted Laser Desorption/Ionization
MCADD	Medium-Chain Acyl-CoA Dehydrogenase Deficiency
MEP	Methylerythritol Phosphate
MGEs	Mobile Genetic Elements
MIRS	Management Information & Retrieval System
MRI	Magnetic Resonance Imaging
mRNA	Messenger RNA

MS	Mass Spectrometry
MVA	Mevalonic Acid
NAD	Nicotinamide Adenine Dinucleotide
NAMAS	National Measurement Accreditation Service
NIRS	Near Infrared Spectroscopy
Nm	Nanometer
NMR	Nuclear Magnetic Resonance
OD	Optical Thickness
PA	Phosphatidic Acid
PC	Phosphatidylcholine
PCA	Principal Components Analysis
PEP	Phosphoenolpyruvate
PG	Phosphatidylglycerol
PGS	Phosphatidyl Glycerophosphate Synthase
PI	Phosphatidylinositol
PKU	Phenylketonuria
PS	Phosphatidylserine
RF	Retardation Factor
RIA	Radioimmunoassay
RM	Restriction-Modification
RNA	Ribonucleic Acid
SA	Sinapinic Acid
SUF	Sulfur Utilization Factor
TOF	Time-Of-Flight
tRNA	Transfer RNA
TSH	Thyroid-Stimulating Hormone
UK NEQAS	United Kingdom National External Quality Assessment Service
UTP	Uridine Triphosphate
UV	Ultra Violet
VOCs	Volatile Organic Compounds
WEQAS	Wales External Quality Assurance Scheme
YFP	Yellow Fluorescent Protein

GLOSSARY

A

Absorption Spectrum- a spectrum of electromagnetic radiation transmitted through a substance, showing dark lines or bands due to absorption at specific wavelengths.

Amniocentesis - Amniocentesis is a medical procedure used primarily in prenatal diagnosis of chromosomal abnormalities and fetal infections as well as for sex determination. In this procedure, a small amount of amniotic fluid, which contains fetal tissues, is sampled from the amniotic sac surrounding a developing fetus

Amphiphilic Nature- A chemical compound possessing both hydrophilic (water-loving, polar) and lipophilic (fat-loving) properties.

Anabolism- the synthesis of complex molecules in living organisms from simpler ones together with the storage of energy.

Assay- An assay is an investigative procedure in laboratory medicine, mining, pharmacology, environmental biology and molecular biology for qualitatively assessing or quantitatively measuring the presence, amount, or functional activity of a target entity.

B

Biomolecule- an organic molecule that includes carbohydrates, protein, lipids, and nucleic acids.

Biopolymers- polymers that are produced by or derived from living organisms,

Biosynthesis- the production of complex molecules within living organisms or cells.

C

Cataract – A clouding or loss of transparency of the lens in the eye as a result of tissue breakdown and protein clumping.

Chirality - In chemistry, a molecule or ion is called chiral if it cannot be superposed on its mirror image by any combination of rotations, translations, and some conformational changes. This geometric property is called chirality.

Chromatographic - In chemical analysis, chromatography is a laboratory technique for the separation of a mixture into its components. The mixture is dissolved in a fluid solvent called the mobile phase, which carries it through a system on which a material called the stationary phase is fixed.

Colorimetry- Is a scientific technique that is used to determine the concentration of colored compounds in solutions by the application of the Beer–Lambert law, which states that the concentration of a solute is proportional to the absorbance.

Condensation - Condensation is the change of the state of matter from the gas phase into the liquid phase, and is the reverse of vaporization.

Cytoplasm - Cytoplasm is a thick solution that fills each cell and is enclosed by the cell membrane. It is mainly composed of water, salts, and proteins. In eukaryotic cells, the cytoplasm includes all of the material inside the cell and outside of the nucleus.

E

Endoproteinases - Endopeptidase or endoproteinase are proteolytic peptidases that break peptide bonds of nonterminal amino acids (i.e., within the molecule), in contrast to exopeptidases, which break peptide bonds from end-pieces of terminal amino acids.
Enzymes- are proteins that help speed up metabolism, or the chemical reactions in our bodies.
Exons- The sequence of DNA present in mature messenger RNA, some of which encodes the amino acids of a protein.

F

Fat - In nutrition, biology, and chemistry, fat usually means any ester of fatty acids, or a mixture of such compounds, most commonly those that occur in living beings or in food.
Fertilization- is the fusion of gametes to give rise to a new individual organism or offspring and initiate its development.
Fluorophores- A fluorophore (or fluorochrome, similarly to a chromophore) is a fluorescent chemical compound that can re-emit light upon light excitation.

G

Glycolipids - Glycolipids are lipids with a carbohydrate attached by a glycosidic (covalent) bond. Their role is to maintain the stability of the cell membrane and to facilitate cellular recognition, which is crucial to the immune response and in the connections that allow cells to connect to one another to form tissues.

H

Hematological - Hematology is the branch of medicine concerned with the study of the cause, prognosis, treatment, and prevention of diseases related to blood.
Hormone - A hormone is any member of a class of signaling molecules in multicellular organisms, which are transported by intricate biological processes to distant organs to regulate physiology and behavior. Hormones are required for the correct development of animals, plants and fungi.

I

Inert Gas - a gas that does not undergo chemical reactions under a set of given conditions.
Introns- are non-coding sections of an RNA transcript, or the DNA encoding it, that are spliced out before the RNA molecule is translated into a protein.

M

Macromolecules- are very large molecules important to biophysical processes, such as a protein or nucleic acid. It is composed of thousands of covalently bonded atoms.

Membranes - Body membranes are thin sheets of tissue that cover the body, line body cavities, and cover organs within the cavities in hollow organs. They can be categorized into epithelial and connective tissue membrane.
Metabolic Pathway- a linked series of chemical reactions occurring within a cell.
Microheterogeneity - Variation in the chemical structure of a substance (as the amino acid sequence of a protein) that does not produce a major change in its properties.
Monochromator - A monochromator is an optical device that transmits a mechanically selectable narrow band of wavelengths of light or other radiation chosen from a wider range of wavelengths available at the input. The name is from the Greek roots mono-, "single", and chroma, "color", and the Latin suffix -ator, denoting an agent.
Mutation- is a change in the DNA sequence of an organism. Mutations can result from errors in DNA replication during cell division, exposure to mutagens or a viral infection.

N

Necrosis - Necrosis is the death of body tissue. It occurs when too little blood flows to the tissue. This can be from injury, radiation, or chemicals.
Neurotransmitter - A neurotransmitter is a signaling molecule secreted by a neuron to affect another cell across a synapse. The cell receiving the signal, any main body part or target cell, may be another neuron, but could also be a gland or muscle cell.
Noncovalent- a type of chemical bond that is found mostly between macromolecules.
Nucleotide- are organic molecules consisting of a nucleoside and a phosphate.

O

Offspring- offspring are the young creation of living organisms, produced either by a single organism or, in the case of sexual reproduction, two organisms.
Oligosaccharides - An oligosaccharide is a saccharide polymer containing a small number of monosaccharides. Oligosaccharides can have many functions including cell recognition and cell binding. For example, glycolipids have an important role in the immune response.
Organelles - Organelles are specialized structures that perform various jobs inside cells. The term literally means "little organs." In the same way organs, such as the heart, liver, stomach, and kidneys, serve specific functions to keep an organism alive, organelles serve specific functions to keep a cell alive.

P

Paracrine - Of or relating to a hormone or to a secretion released by (endocrine) cells into the adjacent cells or surrounding tissue rather than into the bloodstream.
Phenotype- refers to an individual's observable traits, such as height, eye color and blood type.
Phospholipids - Phospholipids are major membrane lipids that consist of lipid bilayers. This basic cellular structure acts as a barrier to protect the cell against various environmental insults and more importantly, enables multiple cellular processes to occur in subcellular compartments.

Protonation- In chemistry, protonation (or hydronation) is the adding of a proton (or hydron, or hydrogen cation), (H+) to an atom, molecule, or ion, forming a conjugate acid.

S

Saccharide - Any of a series of sweet-tasting, crystalline carbohydrates, especially a simple sugar.

Species- a species is the basic unit of classification and a taxonomic rank of an organism, as well as a unit of biodiversity.

Substrates - The substrate is the surface on which an organism grows or is attached in biology. A substrate, for example, could be a microbiological medium. The substrate can also refer to the material found at the bottom of a habitat, such as gravel in an aquarium.

T

Tissue - Tissue is a group of cells that have similar structures and that function together as a unit. A non-living material, called the intercellular matrix, fills the spaces between the cells. This may be abundant in some tissues and minimal in others.

V

Venipuncture - is a surgical puncture of a vein especially for the withdrawal of blood or for intravenous medication.

Volatilization- the conversion of a liquid chemical into a vapor, which escapes into the atmosphere.

PREFACE

This book takes the readers through several aspects of Biochemistry. This book gives an introduction to biochemistry, biosynthesis and immunochemical techniques, genetic information transfer, chromatography and biochemistry, mass Spectrometry and spectroscopic techniques, principles of clinical biochemistry and biochemistry of lipids.

The first chapter stresses the basic introduction to the biochemistry so that the readers gets familiar with the theory and concepts of biochemistry. This chapter will also emphasize on enzymes, carbohydrates, glycobiology, amino acids, peptides, proteins and hormones.

The second chapter takes the readers through the concepts of biosynthesis and immunochemical techniques. This chapter will provide highlights on the biosynthesis of cell constituents, sulfur-containing small biomolecules in plants, volatile plant secondary metabolism, flavonoid biosynthesis and immunochemical techniques.

Then, the third chapter explains genetic information transfer. It also explains the transformation, transduction, conjugation, transcription, translation and steps in translation. It also explains about genetic information in microbes and horizontal gene transfer.

The fourth chapter introduces the readers to chromatographic techniques for analyzing and quantifying biomolecules. This chapter also explains the principles, performance parameters, chromatographic equipment and modes of chromatography. The chapter also sheds light on HPLC, membrane - based chromatography and chromatography of a sample protein.

The fifth chapter throws light on Mass Spectrometry. This chapter contains the definition, principles, and appplication of mass spectroscopy, mass spectrometry of proteins/peptides, fluorescence spectroscopy, fluorescence correlation spectroscopy and infrared spectroscopy.

The sixth chapter takes the readers through concepts of clinical biochemistry. The readers are then told about the clinical biochemical analysis, clinical measurements, quality control and examples of biochemical aids to clinical diagnosis.

The seventh chapter sheds light on spectroscopic techniques in biochemistry. This chapter also mentions various aspects of electromagnetic radiation, interaction with matter, lasers, UV-vis light spectroscopy, its principles, instrumentation and applications. The instrumentation and applications of fluorescence spectroscopy, are explained well in this chapter.

In the last chapter biochemistry of lipids, diversity in lipid structure, properties of lipids in solution, engineering of membrane lipid composition, the role of lipids in cell function, lipid metabolism in plants and plant lipid geography are discussed.

This book has been designed to suit the knowledge and pursuit of the researcher and scholars and to empower them with various aspects of biochemistry so that they are updated with the information. I hope that the readers find the book explanatory and insightful and that this book is referred by scholars across various fields.

CHAPTER 1

INTRODUCTION TO BIOCHEMISTRY

CONTENTS

The study of biochemistry is an essential component to understand the biological process at the cellular and molecular level with the help of chemistry. The field of contemporary biochemistry plays a crucial role in the field of life sciences, whether it is investigating metabolic pathways, storage diseases, mechanism actions of biomolecules, or intra and intercellular communication.

1.1. INTRODUCTION

Enzymes: Body proteins perform a large number of functions. One such unique function is that they act as biological catalysts (Enzymes). They are responsible for highly complex reactions. They direct the metabolic events and exhibit specificity toward substrates, regulating the entire metabolism.

Thus, they play a key role in the degradation and synthesis of nutrients, biomolecules, etc. The most important diagnostic procedures involve the assay of enzymes. As well as determining damaged tissues, the extent of tissue damage, and monitoring the course of the disease, they have therapeutic uses in diagnosing a wide variety of diseases.

Amino acids and Proteins: Living systems are made up of Proteins. They are the dehydration polymers of amino acids. Each amino acid residue is joined by a peptide linkage to form proteins.

In addition to molecular instruments through which genetic information is expressed, proteins make up hormones, antibodies, transporters, the lens protein, the structural framework of our bodies, and a variety of substances with distinct biological functions. The type, nature, and number of amino acids impart characteristic properties to the proteins.

There are about 300 amino acids, but only 20 are coded by the DNA of higher organisms. Acid-base properties of amino acids are important to the individual physical and chemical nature of the protein. The structural organization of proteins could be primary, secondary, tertiary, and quaternary. The three-dimensional structure is the most biologically active one.

The unfolding and disorganization of the proteins results in denaturation, the process is mostly irreversible. Such a protein may lose its biological function. It is known that amino acid-derived peptides play a critical role in the body, and special products made from them are critical to the body.

Carbohydrates: In living organisms, carbohydrates are abundantly found as biomolecules. They contain more than one hydroxyl group (polyhydric) In addition to the aldehyde or ketone group. Thus, they form into polyhydroxy aldoses or polyhydroxy ketoses. Carbohydrates can be classified into monosaccharides, disaccharides, and polysaccharides.

Mono is the smallest sugar unit; disaccharide is made up of two monosaccharides joined by glycosidic linkages. The linkage can be α or β. A polymer with more than 10 monosaccharide units is called a polysaccharide. Carbohydrates have a wide range of functions. They provide energy and act as storage molecules of energy.

They serve as cell membrane components and mediate some forms of communication between cells. The absence of a single enzyme like lactase causes discomfort and diarrhea. The failure of galactose and fructose metabolism due to deficient enzymes leads to the turbidity of lens proteins (Cataract).

Blood glucose is controlled by different hormones and metabolic processes. Diabetes is a condition in which the insulin hormone is insufficient or not functioning properly. These patients are at risk for atherosclerosis, vascular disease, and renal failure.

Lipids: The bulk of the living matter is made up of lipids, carbohydrates, and proteins. Lipids are water-insoluble but can be extracted with non-polar solvents like benzene, methanol, or ether. Some lipids act as storage molecules, for example, triglycerides stored in adipose tissue.

Transport forms of lipids, are present in combination with proteins (Lipoproteins). Building blocks of lipids are fatty acids. Some lipids like cholesterol lack fatty acids but are potentially related to them. Lipids are constituents of cell membranes and act as hydrophobic barriers that permit the entry/exit of certain molecules. Lipids carry fat-soluble vitamins and form special biomolecules.

A lipid imbalance can lead to serious diseases such as obesity and atherosclerosis. Breakdown of fatty acid produces energy; excessive breakdown causes ketosis, ketoacidosis, coma, and death. Cholesterol level in blood is controlled by several regulatory mechanisms. Such information is applied in the treatment of patients with high cholesterol levels.

Vitamins and minerals: Several organic compounds are necessary for the body to function. These compounds include vitamins and minerals. They are not synthesized in the body and need to be provided in the diet. Vitamins do not generate energy. Generally, they are responsible for the maintenance of health and the prevention of chronic diseases.

Grossly there are two groups: Water-soluble vitamins are Vit. B-complex and C. Fat-soluble vitamins are Vit A, D, E, and K. Minerals are elements present in the human body. Elements like C, H, and N are provided by the diet and water. The second group includes Ca, P, Mg, Na, K. Cl, and Sulfur. These are required in large quantities (100mg or more/day).

They are called Macroelements. A third group includes trace elements, which are required in small amounts for example Fe, I, Zn, etc. Fluorine deficiency is associated with tooth decay, excess of it causes fluorosis. Sources and requirements are of physiological importance.

The metabolic role and deficiency disorders are important for the students of health sciences. Vitamins and trace elements are particularly important for patients with gastrointestinal disorders, who are fed on artificial diets or parenteral nutrition.

Hormones: Hormones are chemical messengers secreted by endocrine glands and specific tissues. They reach distant organs and stimulate or inhibit their function. They play an important role in carrying messages to the various organs. They form part of the signaling system.

Hormones are synthesized in one tissue, secreted into the blood, and transported as mobile messengers. When they reach the target tissue, they exhibit their actions. Defects in secretion, function, and metabolism can lead to various diseases.

1.2. THE FOUNDATIONS OF BIOCHEMISTRY

A cataclysmic eruption of energetic subatomic particles caused the universe to appear fifteen to twenty billion years ago. Within seconds, the simplest elements (hydrogen and helium) were formed. As the universe expanded and cooled, material condensed under the influence of gravity to form stars.

Some stars become enormous and then explode as supernovae, releasing the energy needed to fuse simpler atomic nuclei into the more complex elements. Thus, were produced, over billions of years, the Earth itself and the chemical elements found on the Earth today.

Living things appeared around four billion years ago, as simple microorganisms that could extract energy from organic compounds or from sunlight, which they used to produce more complex biomolecules based on simple elements and compounds found on the Earth's surface.

Biochemistry asks how the remarkable properties of living organisms arise from the thousands of different lifeless biomolecules. When these molecules are isolated and examined individually, they conform to all the physical and chemical laws that describe the behavior of inanimate matter—as do all the processes occurring in living organisms.

The study of biochemistry shows how the collections of inanimate molecules that constitute living organisms interact to maintain and perpetuate life animated solely by the physical and chemical laws that govern the nonliving universe.

1.2.1. Cellular Foundations

Even at the cellular level, the unity and diversity of organisms are apparent. The smallest organisms consist of single cells and are microscopic. Larger, multicellular organisms contain many different types of cells, which vary in size, shape, and specialized function.

Despite these obvious differences, all cells of the simplest and most complex organisms share certain fundamental properties, which can be seen at the biochemical level.

***Cells Build Supramolecular Structures*:** Macromolecules and their monomeric subunits differ greatly in size. A molecule of alanine is less than 0.5 nm long. Hemoglobin, the oxygen-carrying protein of erythrocytes (red blood cells), consists of nearly 600 amino acid subunits in four long chains, folded into globular shapes and associated in a structure of 5.5 nm in diameter.

In turn, proteins are much smaller than ribosomes (about 20 nm in diameter), which are in turn much smaller than organelles such as mitochondria, typically 1,000 nm in diameter. It is a long jump from simple molecules to cellular structures that can be seen with a light microscope. The monomeric subunits in proteins, nucleic acids, and polysaccharides are joined by covalent bonds.

In supramolecular complexes, however, macromolecules are held together by noncovalent interactions—much weaker, individually, than covalent bonds.

Among these noncovalent interactions are hydrogen bonds (between polar groups), ionic interactions (between charged groups), hydrophobic interactions (among nonpolar groups in aqueous solution), and Van der Waals interactions—all of which have energies substantially smaller than those of covalent bonds.

The large numbers of weak interactions between macromolecules in supramolecular complexes stabilize these assemblies, producing their unique structures.

1.2.2. Chemical Foundations

Biochemistry aims to explain biological form and function in chemical terms. In the study of living organisms, one of the most fruitful approaches has been the isolation of an individual chemical component, such as a protein, and the analysis of its chemical and structural properties.

By the late eighteenth century, chemists had concluded that the composition of living matter is strikingly different from that of the inanimate world. Antoine Lavoisier (1743–1794) noted the relative chemical simplicity of the "mineral world" and contrasted it with the complexity of the "plant and animal worlds"; the latter, he knew, were composed of compounds rich in the elements carbon, oxygen, nitrogen, and phosphorus.

During the first half of the twentieth century, parallel biochemical investigations of glucose breakdown in yeast and animal muscle cells revealed remarkable chemical similarities in these two very different cell types; the breakdown of glucose in yeast and muscle cells involved the same ten chemical intermediates.

Subsequent studies of many other biochemical processes in many different organisms have confirmed the generality of this observation, neatly summarized by Jacques Monod: "What is true of *E. coli* is true of the elephant."

The current understanding that all organisms share a common evolutionary origin is based in part on this observed universality of chemical intermediates and transformations. Only about 30 of the more than 90 naturally occurring chemical elements are essential to organisms.

Most of the elements in living matter have relatively low atomic numbers; only five have atomic numbers above that of selenium. The four most abundant elements in living organisms, in terms of percentage of a total number of atoms, are hydrogen, oxygen, nitrogen, and carbon, which together make up more than 99% of the mass of most cells.

They are the lightest elements capable of forming one, two, three, and four bonds, respectively; in general, the lightest elements form the strongest bonds. The trace elements represent a minuscule fraction of the weight of the human body, but all are essential to life, usually because they are essential to the function of specific proteins, including enzymes.

Hemoglobin's oxygen-transporting ability, for example, depends entirely on four iron ions that constitute just 0.3% of its mass.

***Macromolecules Are the Major Constituents of Cells*:** Many biological molecules are macromolecules, polymers of high molecular weight assembled from relatively simple precursors. Proteins, nucleic acids, and polysaccharides are produced by the polymerization of relatively small compounds with molecular weights of 500 or less.

The number of polymerized units can range from tens to millions. The synthesis of macromolecules is a major energy-consuming activity of cells. Macromolecules themselves may be further assembled into supramolecular complexes, forming functional units such as ribosomes.

Proteins, long polymers of amino acids, constitute the largest fraction (besides water) of cells. In addition to functioning as catalyzing enzymes, some proteins serve as structural elements, signal receptors, or transporters that transport specific substances into or out of cells. Proteins are perhaps the most versatile of all biomolecules.

The nucleic acids, DNA and RNA, are polymers of nucleotides. They store and transmit genetic information, and some RNA molecules have structural and catalytic roles in supramolecular complexes. The polysaccharides, polymers of simple sugars such as glucose, have two major functions: as energy-yielding fuel stores and as extracellular structural elements with specific binding sites for particular proteins.

Shorter polymers of sugars (oligosaccharides) attached to proteins or lipids at the cell surface serve as specific cellular signals. The lipids, greasy or oily hydrocarbon derivatives, serve as structural components of membranes, energy-rich fuel stores, pigments, and intracellular signals.

In proteins, nucleotides, polysaccharides, and lipids, the number of monomeric subunits is very large: molecular weights in the range of 5,000 to more than 1 million for proteins, up to several billion for nucleic acids, and in the millions for polysaccharides such as starch.

Individual lipid molecules are much smaller (M_r 750 to 1,500) and are not classified as macromolecules. However, large numbers of lipid molecules can associate noncovalently with very large structures. Cellular membranes are built of enormous noncovalent aggregates of lipid and protein molecules.

Proteins and nucleic acids are informational macromolecules: each protein and each nucleic acid has a characteristic information-rich subunit sequence. In many cellular processes, some oligosaccharides, which are chains of six or more different sugars, also serve as points of recognition as they serve as specific points of recognition on the outer surface of cells.

1.2.3. Physical Foundations

Living organisms and cells must perform work to remain alive and reproduce. The reactions that occur within cells, like those that occur in factories, require energy to occur. Energy is also consumed in the motion of a bacterium or an Olympic sprinter, in the flashing of a firefly, or in the electrical discharge of an eel.

And the storage and expression of information require energy, without which structures rich in information inevitably become disordered and meaningless. In the course of evolution, cells have developed highly efficient mechanisms for coupling the energy obtained from sunlight or fuels to the many energy-consuming processes they must carry out.

One goal of biochemistry is to understand, in quantitative and chemical terms, how energy is extracted, channeled, and consumed in living cells. We can consider cellular energy conversions—like all other energy conversions—in the context of the laws of thermodynamics.

There are different kinds and concentrations of molecules and ions within a living organism compared to those in the surrounding environment. A Paramecium in a pond, a shark in the ocean, an erythrocyte in the human bloodstream, an apple tree in an orchard—all are different in composition from their surroundings and, once they have reached maturity, all (to a first approximation) maintain a constant composition in the face of constantly changing surroundings.

Although the characteristic composition of an organism changes little through time, the population of molecules within the organism is far from static. Small molecules, macromolecules, and supramolecular complexes are continuously synthesized and then broken down in chemical reactions that involve a constant flux of mass and energy through the system.

The hemoglobin molecules carrying oxygen from your lungs to your brain at this moment were synthesized within the past month; by next month they will have been degraded and entirely replaced by new hemoglobin molecules.

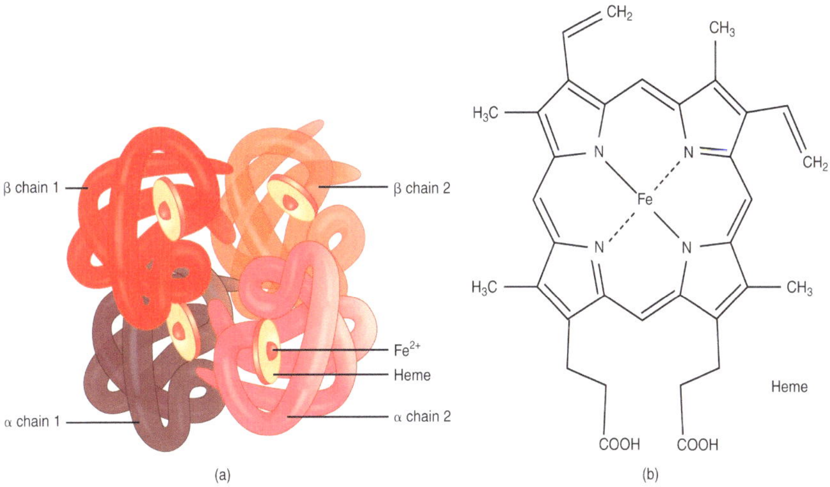

Figure 1.1. Hemoglobin molecules.

Source: Image by Wikimedia Commons

The glucose you ingested with your most recent meal is now circulating in your bloodstream; before the day is over these particular glucose molecules will have been converted into something else—carbon dioxide or fat, perhaps—and will have been replaced with a fresh supply of glucose, so that your blood glucose concentration is more or less constant over the whole day.

The amounts of hemoglobin and glucose in the blood remain nearly constant because the rate of synthesis or intake of each just balances the rate of its breakdown, consumption, or conversion into some other product. The constancy of concentration is the result of a dynamic steady state, a steady-state that is far from equilibrium.

Constant energy investment is necessary to maintain this steady-state; when the cell is no longer able to generate energy, it dies and decays toward equilibrium with the surrounding environment. We consider below exactly what is meant by "steady state" and "equilibrium."

***Enzymes Promote Sequences of chemical reactions*:** Biological macromolecules are much less thermodynamically stable than their monomeric subunits but kinetically stable their uncatalyzed breakdown is so slow (over years rather than seconds) that they are stable on a time scale relevant to the organism.

Virtually every chemical reaction in a cell occurs at a significant rate only because of the presence of enzymes—biocatalysts that, like all other catalysts, greatly enhance the rate of specific chemical reactions without being consumed in the process.

The path from reactant(s) to product(s) almost invariably involves an energy barrier, called the activation barrier that must be surmounted for any reaction to proceed. The breaking of existing bonds and formation of new ones generally requires, first, the distortion of the existing bonds, creating a transition state of higher free energy than either reactant or product.

The highest point in the reaction coordinate diagram represents the transition state, and the difference in energy between the reactant in its ground state and its transition state is the activation energy, $G‡$. During a reaction, enzymes provide a more comfortable fit for the transition state: a surface that is complementary to the transition state in stereochemistry, polarity, and charge.

The binding of enzymes to the transition state is exergonic, and the energy released by this binding reduces the activation energy for the reaction and greatly increases the reaction rate. A further contribution to catalysis occurs when two or more reactants bind to the enzyme's surface close to each other and with stereospecific orientations that favor the reaction.

This increases by orders of magnitude the probability of productive collisions between reactants. Again with a few exceptions, each enzyme catalyzes a specific reaction, and each reaction in a cell is catalyzed by a different enzyme.

Thousands of different enzymes are therefore required by each cell. Having multiple enzymes, specificity (the ability to discriminate between reactants), and susceptibility to regulation, cells can lower activation barriers selectively.

This selectivity is crucial for the effective regulation of cellular processes. By allowing specific reactions to proceed at significant rates at particular times, enzymes determine how matter and energy are channeled into cellular activities.

The thousands of enzyme-catalyzed chemical reactions in cells are functionally organized into many sequences of consecutive reactions, called pathways, in which the product of one reaction becomes the reactant in the next.

Some pathways degrade organic nutrients into simple end products to extract chemical energy and convert it into a form useful to the cell; together these degradative, free-energy-yielding reactions are designated catabolism. Other pathways start with small precursor molecules and convert them to progressively larger and more complex molecules, including proteins and nucleic acids.

Such synthetic pathways, which invariably require the input of energy, are collectively designated anabolism. The overall network of enzyme-catalyzed pathways constitutes cellular metabolism. ATP is the major connecting link (the shared intermediate) between the catabolic and anabolic components of this network.

Proteins, fats, sugars, and nucleic acids are the main constituents of cells, and the pathways of enzyme-catalyzed reactions that act on these components are almost identical in all living organisms.

1.2.4. Genetic Foundations

Living organisms and cells are remarkable for their ability to reproduce themselves for countless generations with excellent fidelity. This continuity of inherited traits implies constancy, over millions of years, in the structure of the molecules that contain the genetic information. Very few historical records of civilization, even those etched in copper or carved in stone, have survived for a thousand years.

But there is good evidence that the genetic instructions in living organisms have remained nearly unchanged over very much longer periods; many bacteria have nearly the same size, shape, and internal structure and contain the same kinds of precursor molecules and enzymes as bacteria that lived nearly four billion years ago.

Among the seminal discoveries in biology in the twentieth century were the chemical nature and the three-dimensional structure of the genetic material, deoxyribonucleic acid, and DNA.

The sequence of the monomeric subunits, the nucleotides (strictly, deoxyribonucleotides), in this linear polymer encodes the instructions for forming all other cellular components and provides a template for the production of identical DNA molecules to be distributed to progeny when a cell divides.

The continued existence of a biological species requires its genetic information to be maintained in a stable form, expressed accurately in the form of gene products, and reproduced with a minimum of errors. Individual species are defined, distinguished, and preserved over successive generations by the way they store, express, and reproduce their genetic messages.

1.2.5. Evolutionary Foundations

Great progress in biochemistry and molecular biology during the decades since Dobzhansky made this striking generalization has amply confirmed its validity.

The astonishing similarity in metabolic pathways and gene sequences across species strongly suggests that all modern organisms share a common evolutionary progenitor and that they were derived from it by a series of small changes (mutations) that conferred selective advantages to some organisms in specific ecological niches.

***Changes in the Hereditary Instructions Allow Evolution*:** While genetic replication is nearly fidelity-perfect, infrequent, unrepaired errors in the replication process lead to alterations in nucleotide sequences, producing genetic mutations, and altering cellular instructions.

Incorrectly repaired damage to one of the DNA strands has the same effect. Mutations in the DNA handed down to offspring—that is, mutations that are carried in the reproductive cells—may be harmful or even lethal to the organism; they may, for example, cause the synthesis of a defective enzyme that is not able to catalyze an essential metabolic reaction.

Occasionally, however, a mutation better equips an organism or cell to survive in its environment. The mutant enzyme might have acquired a slightly different specificity, for example, so that it is now able to use some compound that the cell was previously unable to metabolize.

If a population of cells was to find itself in an environment where that compound was the only or the most abundant available source of fuel, the mutant cell would have a selective advantage over the other, unmutated (wild-type) cells in the population.

The mutant cell and its progeny would survive and prosper in the new environment, whereas wild-type cells would starve and be eliminated. This is what Darwin meant by "survival of the fittest under selective pressure." Occasionally, a whole gene is duplicated.

By producing a new gene with a new function and maintaining the original gene and gene function, the second copy of this gene will not be deleterious; it becomes a means of evolution. Seen in this light, the DNA molecules of modern organisms are historical documents, records of the long journey from the earliest cells to modern organisms.

The historical accounts of DNA are not complete; in the course of evolution, many mutations must have been erased or written over. But DNA molecules are the best source of biological history that we have.

Several billion years of adaptive selection have refined cellular systems to take maximum advantage of the chemical and physical properties of the molecular raw materials for carrying out the basic energy-transforming and self-replicating activities of a living cell.

Chance genetic variations in individual members of a population, coupled with natural selection (survival and reproduction of the fittest individuals in a challenging or shifting environment), have led to the evolution of a wide range of organisms, each adapted to its ecological niche.

***Biomolecules First Arose by Chemical Evolution*:** In our account thus far, we have passed over the first chapter of the story of evolution: the appearance of the first living cell. As well as being present in living organisms, organic compounds such as amino acids and carbohydrates are only present in trace amounts in the earth's crust, in the sea, and the atmosphere.

How did the first living organisms acquire their characteristic organic building blocks? In 1922, the biochemist Aleksandr I. Oparin proposed a theory for the origin of life early in the history of Earth, postulating that the atmosphere was very different from that of today. Rich in methane, ammonia, and water, and essentially devoid of oxygen, it was a reducing atmosphere, in contrast to the oxidizing environment of our era.

In Oparin's theory, electrical energy from lightning discharges or heat energy from volcanoes caused ammonia, methane, water vapor, and other components of the primitive atmosphere to react, forming simple organic compounds. These compounds then dissolved in the ancient seas, which over many millennia became enriched with a large variety of simple organic substances.

In this warm solution (the "primordial soup"), some organic molecules had a greater tendency than others to associate with larger complexes. Over millions of years, these, in turn, assembled spontaneously to form membranes and catalysts (enzymes), which came together to become precursors of the earliest cells.

Oparin's views remained speculative for many years and appeared untestable—until a surprising experiment was conducted using simple equipment on a desktop.

1.3. AN INTRODUCTION TO ENZYMES

There are two fundamental conditions for life. For a living entity to be successful, it must be able to self-replicate. Secondly, it must be able to catalyze chemical reactions efficiently and selectively. The central importance of catalysis may surprise some beginning students of biochemistry, but it is easy to demonstrate.

Many of us, for example, consume substantial amounts of sucrose—common table sugar—as a kind of fuel, whether in the form of sweetened foods and drinks or as sugar itself. The conversion of sucrose to CO_2 and H_2O in the presence of oxygen is a highly exergonic process, releasing free energy that we can use to think, move, taste, and see.

However, a bag of sugar can remain on the shelf for years without any obvious conversion to CO_2 and H_2O. Although this chemical process is thermodynamically favorable, it is very slow! Yet when sucrose is consumed by a human (or almost any other organism), it releases its chemical energy in seconds.

The difference is catalysis. Without catalysis, chemical reactions such as sucrose oxidation could not occur on a useful time scale and thus could not sustain life.

The catalytic power of enzymes is often much greater than that of synthetic or inorganic catalysts. They have a high degree of specificity for their substrates, they accelerate chemical reactions tremendously, and they function in aqueous solutions under very mild conditions of temperature and pH. Few nonbiological catalysts have all these properties.

Enzymes are central to every biochemical process. Acting in organized sequences, they catalyze the hundreds of stepwise reactions that degrade nutrient molecules, conserve and transform chemical energy, and make biological macromolecules from simple precursors.

Through the action of regulatory enzymes, metabolic pathways are highly coordinated to yield a harmonious interplay among the many activities necessary to sustain life. The study of enzymes has immense practical importance. In some diseases, especially inheritable genetic disorders, there may be a deficiency or even a total absence of one or more enzymes.

For other disease conditions, excessive activity of an enzyme may be the cause. It is important to measure the enzyme activity in blood plasma, erythrocytes, or tissue samples to diagnose certain illnesses. Many drugs exert their biological effects through interactions with enzymes.

And enzymes are important practical tools, not only in medicine but in the chemical industry, food processing, and agriculture. We begin with descriptions of the properties of enzymes and the principles underlying their catalytic power, and then introduce enzyme kinetics, a discipline that provides much of the framework for any discussion of enzymes.

Specific examples of enzyme mechanisms are then provided, illustrating principles introduced earlier in the chapter. We end with a discussion of how enzyme activity is regulated.

Much of the history of biochemistry is the history of enzyme research. In the late 1700s, studies on the digestion of meat by secretions of the stomach first described biological catalysis, and research continued in the 1800s with studies on how starch is converted to sugar by saliva and various plant extracts.

In the 1850s, Louis Pasteur concluded that the fermentation of sugar into alcohol by yeast is catalyzed by "ferments." He postulated that these ferments were inseparable from the structure of living yeast cells; this view, called vitalism, prevailed for decades.

Then in 1897, Eduard Buchner discovered that yeast extracts could ferment sugar to alcohol, proving that fermentation was promoted by molecules that continued to function when removed from cells. Frederick W. Kühne called these molecules enzymes. As vitalistic notions of life were disproved, the isolation of new enzymes and the investigation of their properties advanced the science of biochemistry.

The isolation and crystallization of urease by James Sumner in 1926 provided a breakthrough in early enzyme studies. Sumner found that urease crystals consisted entirely of protein, and he postulated that all enzymes are proteins.

In the absence of other examples, this idea remained controversial for some time. Only in the 1930s was Sumner's conclusion widely accepted, after John Northrop and Moses Kunitz crystallized pepsin, trypsin, and other digestive enzymes and found them also to be proteins. During this period, J. B. S. Haldane wrote a treatise entitled Enzymes.

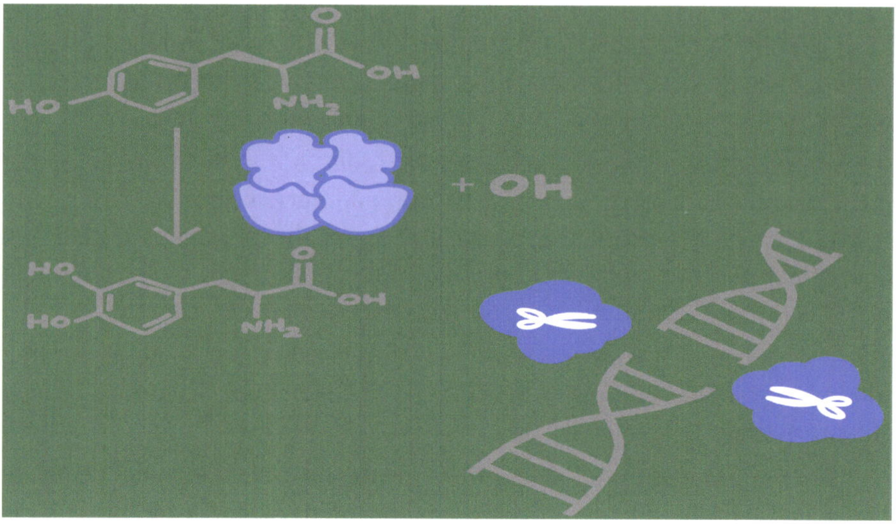

Figure 1.2. Enzyme.

Source: Image by Innovative Genomics Institute

Although the molecular nature of enzymes was not yet fully appreciated, Haldane made the remarkable suggestion that weak bonding interactions between an enzyme and its substrate might be used to catalyze a reaction. This insight lies at the heart of our current understanding of enzymatic catalysis. Since the latter part of the twentieth century, research on enzymes has been intensive.

As a result, thousands of enzymes have been purified, the structure and chemical mechanism of many of them have been revealed, and a general understanding of how enzymes function has been gained.

Enzymes are protein catalysts for chemical reactions in biological systems. They increase the rate of chemical reactions taking place within living cells without changing themselves. Most enzymes are protein in nature. Depending on the presence and absence of a nonprotein component with the enzyme enzymes can exist as, simple enzymes or holoenzymes:

- Simple enzyme: It is made up of only protein molecules not bound to any nonproteins. Example: Pancreatic Ribonuclease.
- Holo enzymes are made up of protein groups and non-protein components.

- The protein component of this holoenzyme is called apoenzyme.
- The non-protein component of the holoenzyme is called a cofactor.

If this cofactor is an organic compound it is called a coenzyme and if it is an inorganic group it is called an activator (Fe 2+, Mn 2+, or Zn 2+ ions). If the cofactor is bound so tightly to the apoenzyme and is difficult to remove without damaging the enzyme it is sometimes called a prosthetic group.

1.3.1. Coenzymes

- Coenzymes are derivatives of vitamins without which the enzyme cannot exhibit any reaction. One molecule of the coenzyme is able to convert a large number of substrate molecules with the help of enzymes.
- Coenzyme accepts a particular group removed from the substrate or donates a particular group to the substrate.
- Coenzymes are called co-substrates because the changes that take place in substrates are complementary to the changes in coenzymes.
- The coenzyme may participate in forming an intermediate enzyme-substrate complex Example: NAD, FAD, Coenzyme A

Many enzymes require metal ions like Ca2+, K+, Mg2+, Fe2+, Cu2+, Zn2+, Mn2+, and Co2+ for their activity. Metal-activated enzymes form only loose and easily dissociable complexes with the metal and can easily release the metal without denaturation. Metalloenzymes hold the metal tightly on the molecule and do not release it even during extensive purification. Metal ions promote enzyme action by:

- Maintaining or producing the active structural conformation of the enzyme (e.g., glutamine synthase)
- Promoting the formation of the enzyme-substrate complex (Example: Enolase and carboxypeptidase A.)
- Acting as electron donors or acceptors (Example: Fe-S proteins and cytochromes)
- Causing distortions in the substrate or the enzyme Example: phosphotransferases).

1.3.2. Properties of Enzymes

Active site: Enzyme molecules contain a special pocket or cleft called the active site. The active site contains amino acid chains that create a three-dimensional surface complementary to the substrate. The active site binds the substrate, forming an enzyme-substrate (ES) complex.

ES is converted to enzyme-product (EP); which subsequently dissociates to enzyme and product. For the combination with the substrate, each enzyme is said to possess one or more active sites where the substrate can be taken up.

The active site of the enzyme may contain a free hydroxyl group of serine, phenolic (hydroxyl) group of tyrosine, SH-thiol (Sulfhydryl) group of cysteine, or imidazole group of histidine to interact with there is substrates. It is also possible that the active site (Catalytic site) is different from the binding site in which case they are situated closely together in the enzyme molecule.

***Catalytic efficiency/ Enzyme turnover number*:** Most enzyme-catalyzed reactions are highly efficient proceeding from 103 to 108 times faster than uncatalyzed reactions. Typically, each enzyme molecule is capable of transforming 100 to 1000 substrate molecules into products each second.

Enzyme turnover number refers to the amount of substrate converted per unit time (carbonic anhydrase is the fastest enzyme).

***Specificity*:** Each enzyme has a specific substrate. The specificity of enzymes is divided into a. Absolute specificity:- this means one enzyme catalyzes or acts on only one substrate.

For example, Urease catalyzes the hydrolysis of urea but not thiourea. b. Stereo specificity- some enzymes catalyze only one type of molecule, even if the compound is one type of molecule: glucose oxidase catalyzes the oxidation of β-D-glucose but not α-D-glucose, and arginase catalyzes the hydrolysis of L-arginine but not D-arginine. Maltase catalyzes the hydrolysis of α- but not β –glycosides.

Zymogens (- an inactive form of enzyme): Some enzymes are produced in nature in an inactive form that can be activated when they are required. Such types of enzymes are called Zymogens (Proenzymes). Many of the digestive enzymes and enzymes concerned with blood coagulation are in this group Examples: Pepsinogen - This zymogen is from gastric juice.

When required Pepsinogen converts to Pepsin Trypsinogen - This zymogen is found in the pancreatic juice, and when it is required gets converted to trypsin. Zymogen forms enzymes, a protective mechanism to prevent auto digestion of tissue producing the digestive enzymes and preventing intravascular coagulation of blood.

Isoenzymes (Isozymes): This group of enzymes possesses similar catalytic activity, acts on similar substrates, and produces similar products, but originated at different sites and exhibit different physical and chemical properties, including electrophoretic mobilities, amino acid composition, and immunological behavior.

1.3.3. Classification of Enzymes

Enzymes are classified based on the reactions they catalyze- Each enzyme is assigned a four-digit classification number and a systematic name, which identifies the reaction catalyzed. The international union of Biochemistry and Molecular Biology developed a system of nomenclature on which enzymes are divided into six major classes, each with numerous subgroups.

Enzymes are classified based on the reactions they catalyze. Each enzyme is characterized by a code number comprising four digits separated by points. The four digits characterize the class, sub-class, sub-sub-class, and a serial number of a particular enzyme.

1.3.4. Mechanism of Enzyme Action (1913)

Michaels and Menten have proposed a hypothesis for enzyme action, which is most acceptable. Based on their hypothesis, the enzyme molecule (E) first combines with a substrate molecule (S) to form an enzyme-substrate (ES) complex, which then dissociates into the product (P) and enzyme (E).

Enzymes once dissociated from the complex are free to combine with another molecule of the substrate and similarly form a product.

It is possible for a chemical reaction S->P (where S is the substrate and P is the product or products) to occur if a certain number of molecules in S are in an activated state called the "transition state," in which there is a high probability of forming or breaking a chemical bond. The transition state is the top of the energy barrier separating the reactants and products.

The rate of a given reaction will vary directly as the number of reactant molecules in the transition state. The "energy of activation is the amount of energy required to bring all the molecules in 1 gram-mole of a substrate at a given temperature to the transition state A temperature rise, by increasing thermal motion and energy, causes an increase in the number of molecules on the transition state and thus accelerates a chemical reaction.

The addition of an enzyme or any catalyst can also bring about such acceleration. The enzyme combines transiently with the substrate to produce a transient state having a lower energy of activation than that of the substrate alone. This results in the acceleration of the reaction.

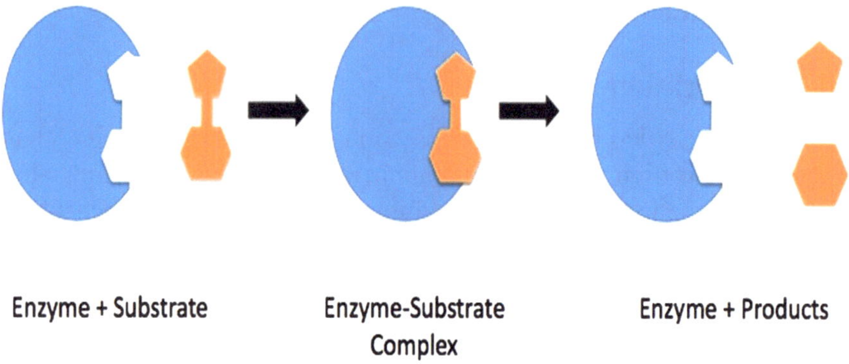

Figure 1.3. Enzyme mechanism- a model.

Source: Image by Wikimedia Commons

Once the products are formed, the enzyme (or catalyst) is free or regenerated to combine with another molecule of the substrate and repeat the process. Activation energy is defined as the energy required for converting all molecules in one mole of a reacting substance from the ground state to the transition state. Enzymes are said to reduce the magnitude of this activation energy.

1.4. CARBOHYDRATES AND GLYCOBIOLOGY

Carbohydrates are the most abundant biomolecules on Earth. The photosynthesis process converts more than 100 billion metric tons of CO^2 and H^2O into cellulose and other plant products every year. Certain carbohydrates (sugar and starch) are a dietary staple in most parts of the world, and the oxidation of carbohydrates is the central energy-yielding pathway in most non-photosynthetic cells.

Insoluble carbohydrate polymers serve as structural and protective elements in the cell walls of bacteria and plants and the connective tissues of animals. Other carbohydrate polymers lubricate skeletal joints and participate in recognition and adhesion between cells.

More complex carbohydrate polymers covalently attached to proteins or lipids act as signals that determine the intracellular location or metabolic fate of these hybrid molecules, called glycoconjugates. This chapter introduces the major classes of carbohydrates and glycoconjugates and provides a few examples of their many structural and functional roles.

Carbohydrates are polyhydroxy aldehydes or ketones, or substances that yield such compounds on hydrolysis. Many, but not all, carbohydrates have the empirical formula $(CH_2O)_n$; some also contain nitrogen, phosphorus, or sulfur.

There are three major size classes of carbohydrates: monosaccharides, oligosaccharides, and polysaccharides (the word "saccharide" is derived from the Greek sakcharon, meaning "sugar"). Monosaccharides, or simple sugars, consist of a single polyhydroxy aldehyde or ketone unit.

The most abundant monosaccharide in nature is the six-carbon sugar D-glucose, sometimes referred to as dextrose. Monosaccharides of more than four carbons tend to have cyclic structures. Oligosaccharides consist of short chains of monosaccharide units, or residues, joined by characteristic linkages called glycosidic bonds.

The most abundant are the disaccharides, with two monosaccharide units. Typical is sucrose (cane sugar), which consists of the six-carbon sugars D-glucose and D-fructose. All common monosaccharides and disaccharides have names ending with the suffix "-ose."

In cells, most oligosaccharides consisting of three or more units do not occur as free entities but are joined to non-sugar molecules (lipids or proteins) in glycoconjugates. The polysaccharides are sugar polymers containing more than 20 or so monosaccharide units, and some have hundreds or thousands of units. Some polysaccharides, such as cellulose, are linear chains; others, such as glycogen, are branched.

Despite both having recurring units of D-glucose, glycogen and cellulose differ in the type of glycosidic linkage they possess, so their properties and biological functions are strikingly different.

1.4.1. Monosaccharides and Disaccharides

The simplest carbohydrate molecules are either aldehydes or ketones that contain at least two hydroxyl groups; glucose and fructose, which are six-carbon monosaccharides, have five hydroxyl groups each. Many of the carbon atoms to which hydroxyl groups are attached are chiral centers, which give rise to the many sugar stereoisomers found in nature.

We begin by describing the families of monosaccharides with backbones of three to seven carbons—their structure and stereoisomeric forms, and the means of representing their three-dimensional structures on paper. We then discuss several chemical reactions of the carbonyl groups of monosaccharides.

One such reaction, the addition of a hydroxyl group from within the same molecule, generates the cyclic forms of five- and six-carbon sugars (the forms that predominate in aqueous solution) and creates a new chiral center, adding further stereochemical complexity to this class of compounds.

1.4.2. The Two Families of Monosaccharides Are Aldoses and Ketoses

Monosaccharides are colorless, crystalline solids that are freely soluble in water but insoluble in nonpolar solvents. Most have a sweet taste.

The backbones of common monosaccharide molecules are unbranched carbon chains in which all the carbon atoms are linked by single bonds. In the open-chain form, one of the carbon atoms is double-bonded to an oxygen atom to form a carbonyl group; each of the other carbon atoms has a hydroxyl group.

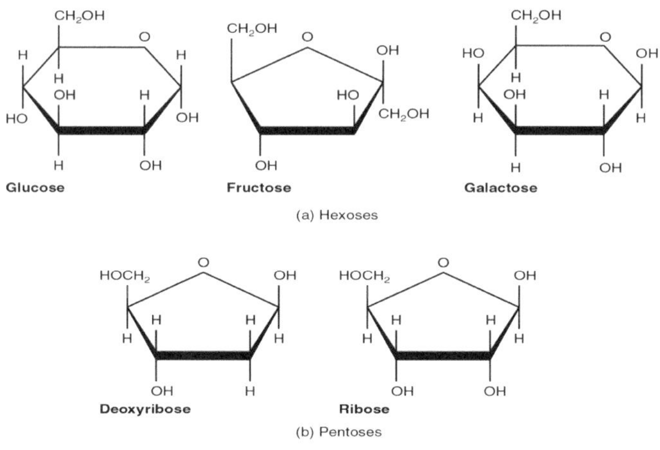

Figure 1.4. Five Important Monosaccharides.

Source: Image by Wikimedia Commons

If the carbonyl group is at an end of the carbon chain (that is, in an aldehyde group) the monosaccharide is an aldose; if the carbonyl group is at any other position (in a ketone group) the monosaccharide is a ketose. The simplest monosaccharides are the two three-carbon trioses: glyceraldehyde, an aldo-triose, and dihydroxyacetone, a keto-triose.

Monosaccharides with four, five, six, and seven carbon atoms in their backbones are called, respectively, tetroses, pentoses, hexoses, and heptoses. There are aldoses and ketoses of each of these chain lengths: aldotetroses and ketotetrose, aldopentoses and ketopentoses, and so on.

The hexoses, which include the aldohexose D-glucose and the ketohexose D-fructose, are the most common monosaccharides in nature. The aldopentoses D-ribose and 2-deoxy-D-ribose are components of nucleotides and nucleic acids.

1.4.3. Polysaccharides

The majority of carbohydrates found in nature are polysaccharides, polymers with medium to high molecular weights. Glycans, such as polysaccharides, differ from one another in terms of their recurring monosaccharide units, the length of their chains, the types of bonds linking the units, and the degree of branching.

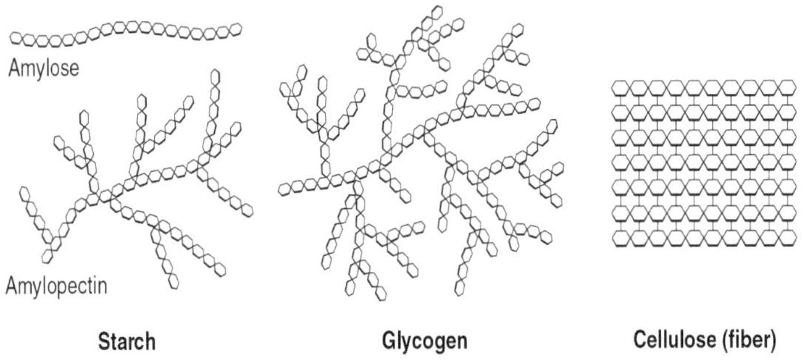

Figure 1.5. Three Important Polysaccharides.

Source: Image by Wikimedia Commons

Homopolysaccharides contain only a single type of monomer; heteropolysaccharides contain two or more different kinds. Some homopolysaccharides serve as storage forms of monosaccharides that are used as fuels; starch and glycogen are homopolysaccharides of this type.

Other homopolysaccharides (cellulose and chitin, for example) serve as structural elements in plant cell walls and animal exoskeletons.

Heteropolysaccharides provide extracellular support for organisms of all kingdoms. For example, the rigid layer of the bacterial cell envelope (the peptidoglycan) is composed of part of a heteropolysaccharide built from two alternating monosaccharide units.

In animal tissues, the extracellular space is occupied by several types of heteropolysaccharides, which form a matrix that holds individual cells together and provides protection, shape, and support to cells, tissues, and organs.

Unlike proteins, polysaccharides generally do not have definite molecular weights. This difference is a consequence of the mechanisms of assembly of the two types of polymers. Proteins are synthesized on a template (messenger RNA) of defined sequence and length, by enzymes that follow the template exactly.

For polysaccharide synthesis there is no template; rather, the program for polysaccharide synthesis is intrinsic to the enzymes that catalyze the polymerization of the monomeric units, and there is no specific stopping point in the synthetic process.

1.4.4. Some Homopolysaccharides Are Stored Forms of Fuel

In plant cells, starch is the most important storage polysaccharide, and in animal cells, glycogen. Both polysaccharides occur intracellularly as large clusters or granules. Starch and glycogen molecules are heavily hydrated because they have many exposed hydroxyl groups available to hydrogen-bond with water.

Most plant cells can form starch, but it is especially abundant in tubers, such as potatoes, and seeds. Starch contains two types of glucose polymer, amylose, and amylopectin. The former consists of long, unbranched chains of D-glucose residues connected by ($^{1}n_{4}$) linkages. Such chains vary in molecular weight from a few thousand to more than a million.

Amylopectin also has a high molecular weight (up to 100 million) but unlike amylose is highly branched. The glycosidic linkages joining successive glucose residues in amylopectin chains are ($^{1}n_{4}$); the branch points (occurring every 24 to 30 residues) are ($^{1}n_{6}$) linkages.

Glycogen is the main storage polysaccharide of animal cells. Like amylopectin, glycogen is a polymer of ($^{1}n_{4}$)-linked subunits of glucose, with ($^{1}n_{6}$)-linked branches, but glycogen is more extensively branched (on average, every 8 to 12 residues) and more compact than starch.

Glycogen is especially abundant in the liver, where it may constitute as much as 7% of the wet weight; it is also present in skeletal muscle. In hepatocytes, glycogen is found in large granules, which are themselves clusters of smaller granules composed of single, highly branched glycogen molecules with an average molecular weight of several million.

Such glycogen granules also contain, in tightly bound form, the enzymes responsible for the synthesis and degradation of glycogen. Because each branch in glycogen ends with a nonreducing sugar unit, a glycogen molecule has as many nonreducing ends as it has branches, but only one reducing end.

When glycogen is used as an energy source, glucose units are removed one at a time from the nonreducing ends. Branches with nonreducing ends can be attacked simultaneously by enzymes that act only on nonreducing ends, speeding up the conversion of polymers into monosaccharides. Why not store glucose in its monomeric form?

It has been calculated that hepatocytes store glycogen equivalent to a glucose concentration of 0.4 M. The actual concentration of glycogen, which is insoluble and contributes little to the osmolarity of the cytosol, is about 0.01 M. If the cytosol contained 0.4 M glucose, the osmolarity would be threateningly elevated, leading to the osmotic entry of water that might rupture the cell.

Furthermore, with an intracellular glucose concentration of 0.4 M and an external concentration of about 5 mM (the concentration in the blood of a mammal), the free-energy change for glucose uptake into cells against this very high concentration gradient would be prohibitively large.

Dextrans are bacterial and yeast polysaccharides made up of ($^{1}n_{6}$)-linked poly-D-glucose; all have ($^{1}n_{3}$) branches, and some also have ($^{1}n_{2}$) or ($^{1}n_{4}$) branches. Dental plaque, formed by bacteria growing on the surface of teeth, is rich in dextrans.

Synthetic dextrans are used in several commercial products (for example, Sephadex) that serve in the fractionation of proteins by size-exclusion chromatography. Chemically cross-linked dextrans in these products produce insoluble materials of various porosities that can admit macromolecules of various sizes.

1.5. AMINO ACIDS, PEPTIDES, AND PROTEINS

Cells and all parts of cells contain proteins, which are the most abundant macromolecules in nature. Proteins also occur in great variety; thousands of different kinds, ranging in size from relatively small peptides to huge polymers with molecular weights in the millions, may be found in a single cell.

Moreover, proteins exhibit enormous diversity of biological functions and are the most important final products of the information pathways. Proteins are the molecular instruments through which genetic information is expressed. Relatively simple monomeric subunits provide the key to the structure of the thousands of different proteins.

All proteins, whether from the most ancient lines of bacteria or the most complex forms of life, are constructed from the same ubiquitous set of 20 amino acids, covalently linked in characteristic linear sequences.

Because each of these amino acids has a side chain with distinctive chemical properties, this group of 20 precursor molecules may be regarded as the alphabet in which the language of protein structure is written. What is most remarkable is that cells can produce proteins with strikingly different properties and activities by joining the same 20 amino acids in many different combinations and sequences.

From these building blocks, different organisms can make such widely diverse products as enzymes, hormones, antibodies, transporters, muscle fibers, the lens protein of the eye, feathers, spider webs, rhinoceros horn, milk proteins, antibiotics, mushroom poisons, and myriad other substances having distinct biological activities.

Among these protein products, the enzymes are the most varied and specialized. Virtually all cellular reactions are catalyzed by enzymes. Protein structure and function are the topics of this and the next three chapters. We begin with a description of the fundamental chemical properties of amino acids, peptides, and proteins.

1.5.1. Amino Acids

Proteins are polymers of amino acids, with each amino acid residue joined to its neighbor by a specific type of covalent bond. (The term "residue" reflects the loss of the elements of water when one amino acid is joined to another.)

Proteins can be broken down (hydrolyzed) to their constituent amino acids by a variety of methods, and the earliest studies of proteins naturally focused on the free amino acids derived from them. Twenty different amino acids are commonly found in proteins.

The first to be discovered was asparagine, in 1806. The last of the 20 to be found, threonine, was not identified until 1938. All the amino acids have trivial or common names, in some cases derived from the source from which they were first isolated.

Asparagine was first found in asparagus, and glutamate in wheat gluten; tyrosine was first isolated from cheese (its name is derived from the Greek tyros, "cheese"); and glycine (Greek glykos, "sweet") was so named because of its sweet taste.

All 20 of the common amino acids are -amino acids. They have a carboxyl group and an amino group bonded to the same carbon atom (the carbon). They differ from each other in their side chains, or R groups, which vary in structure, size, and electric charge, and which influence the solubility of the amino acids in water.

In addition to these 20 amino acids, there are many fewer common ones. Some are residues modified after a protein has been synthesized; others are amino acids present in living organisms but not as constituents of proteins.

As shorthand for indicating the composition and sequence of amino acids polymerized in proteins, three-letter abbreviations and one-letter symbols have been assigned to the amino acids found in proteins.

1.5.2. Peptides and Proteins

We now turn to polymers of amino acids, peptides, and proteins. The size and number of amino acid residues in biological polypeptides can vary from two or three to thousands. Our focus is on the fundamental chemical properties of these polymers.

Two amino acid molecules can be covalently joined through a substituted amide linkage, termed a peptide bond, to yield a dipeptide. Such a linkage is formed by the removal of the elements of water (dehydration) from the -carboxyl group of one amino acid and the -amino group of another. Peptide bond formation is an example of a condensation reaction, a common class of reactions in living cells.

To make the reaction thermodynamically more favorable, the carboxyl group must be chemically modified or activated so that the hydroxyl group can be more readily eliminated. Three amino acids can be joined by two peptide bonds to form a tripeptide; similarly, amino acids can be linked to form tetrapeptides, pentapeptides, and so forth.

When a few amino acids are joined in this fashion, the structure is called an oligopeptide. When many amino acids are joined, the product is called a polypeptide. Proteins may have thousands of amino acid residues.

Although the terms "protein" and "polypeptide" are sometimes used interchangeably, molecules referred to as polypeptides generally have molecular weights below 10,000, and those called proteins have higher molecular weights.

One can make generalizations about the molecular weights of biologically active peptides and proteins based on their functions. Naturally occurring peptides range in length from two to many thousands of amino acid residues. Even the smallest peptides can have biologically important effects.

Consider the commercially synthesized dipeptide L-aspartyl-L-phenylalanine methyl ester, the artificial sweetener better known as aspartame or NutraSweet. Many small peptides exert their effects at very low concentrations. For example, several vertebrate hormones are small peptides.

These include oxytocin (nine amino acid residues), which is secreted by the posterior pituitary and stimulates uterine contractions; bradykinin (nine residues), which inhibits inflammation of tissues; and thyrotropin-releasing factor (three residues), which is formed in the hypothalamus and stimulates the release of another hormone, thyrotropin, from the anterior pituitary gland.

Some extremely toxic mushroom poisons, such as amanitin, are also small peptides, as are many antibiotics. Slightly larger are small polypeptides and oligopeptides such as the pancreatic hormone insulin, which contains two polypeptide chains, one having 30 amino acid residues and the other 21.

Glucagon, another pancreatic hormone, has 29 residues; it opposes the action of insulin. Corticotropin is a 39-residue hormone of the anterior pituitary gland that stimulates the adrenal cortex. How long are the polypeptide chains in proteins?

Human cytochrome c has 104 amino acid residues linked in a single chain; bovine chymotrypsinogen has 245 residues. At the extreme is titin, a constituent of vertebrate muscle, which has nearly 27,000 amino acid residues and a molecular weight of about 3,000,000. The vast majority of naturally occurring proteins are much smaller than this, containing fewer than 2,000 amino acid residues.

A single polypeptide chain can make up some proteins, but others, called multisub-unit proteins, consist of multiple polypeptide chains that are noncovalently linked. The individual polypeptide chains in a multi-subunit protein may be identical or different. If at least two are identical the protein is said to be oligomeric, and the identical units (consisting of one or more polypeptide chains) are referred to as protomers.

1.5.3. Protein Sequences and Evolution

The simple string of letters denoting the amino acid sequence of a given protein belies the wealth of information this sequence holds. As more protein sequences have become available, the development of more powerful methods for extracting information from them has become a major biochemical enterprise.

Each protein's function relies on its three-dimensional structure, which in turn is determined largely by its primary structure. Thus, the biochemical information conveyed by a protein sequence is in principle limited only by our understanding of structural and functional principles.

On a different level of inquiry, protein sequences are beginning to tell us how the proteins evolved and, ultimately, how life evolved on this planet.

Emile Zuckerkandl and Linus Pauling were pioneers in using nucleotide and protein sequences to study evolution in the mid-1960s, which led to a new field known as molecular evolution. The premise is deceptively straightforward.

If two organisms are closely related, the sequences of their genes and proteins should be similar. The sequences increasingly diverge as the evolutionary distance between two organisms increases. The promise of this approach began to be realized in the 1970s when Carl Woese used ribosomal RNA sequences to define archaebacteria as a group of living organisms distinct from other bacteria and eukaryotes.

Protein sequences offer an opportunity to greatly refine the available information. With the advent of genome projects investigating organisms from bacteria to humans, the number of available sequences is growing at an enormous rate. This information can be used to trace biological history. The challenge is in learning to read the genetic hieroglyphics.

Evolution has not taken a simple linear path. Complexities abound in any attempt to mine the evolutionary information stored in protein sequences. For a given protein, the amino acid residues essential for the activity of the protein are conserved over evolutionary time.

The residues that are less important to function may vary over time—that is, one amino acid may substitute for another—and these variable residues can provide the information used to trace evolution. Amino acid substitutions are not always random, however.

At some positions in the primary structure, the need to maintain protein function may mean that only particular amino acid substitutions can be tolerated. Some proteins have more variable amino acid residues than others. For these and other reasons, proteins can evolve at different rates.

1.6. HORMONES

Hormones are responsible for monitoring changes in the internal and external environment. They direct the body to make necessary adaptations to these environmental changes. Hormones are also produced ectopically by malignant tumors. Tissue production (paracrine) of hormones is also possible. Hormones and the Central nervous system interact to shape development, physiology, behavior, and cognition.

The actions and interactions of the endocrine and nervous systems control neurological activities as well as endocrine functions. Interaction is required for cell-to-cell communication. A messenger secreted by neurons is a neurotransmitter while the secretion of the endocrine is called hormone.

Cellular functions are regulated by hormones, neurotransmitters, and growth factors through their interaction with the receptors, located at the cell surface. Some hormones elicit a hormonal cascade system. A part of the chapter discusses receptors, signal transduction, and second messenger pathways.

Finally, oncogenes and receptor functions are presented. Both hyper and hypo-function of the endocrine glands produce distinct clinical symptoms. The basic information serves as a solid foundation from which to view the existing and future developments in the rapidly advancing discipline.

Hormones are chemical messengers secreted into the blood by endocrine or ductless glands. However, many hormones are secreted by organs that are not ductless glands. Hormone means to arouse or excite. Major endocrine glands are the pituitary, hypothalamus, thyroid; adrenals, pancreas, ovaries, and testes. Others are Thymus, Pineal gland, and gastrointestinal hormones. Hormones can be classified based on their structure, mechanism of action, based on their site of production, etc. Hormones reach target organs, exert their metabolic effects, and also reach their site of production.

Here, they inhibit the production of thc hormonc. This is called feedback inhibition. Sometimes the concentration of the hormone is less, which stimulates the production of hormone by a process of feedback stimulation.

1.6.1. Biosynthesis of Hormones

Biosynthetic mechanisms are many. It is necessary to remove certain peptide sequences from some protein hormone precursors in order to convert them into their active form. E.g., Insulin is synthesized as pre-proinsulin (m.wt-11500). Removal of some amino acids and peptides produces insulin (m.wt-5734).

Thyroxine is a single amino acid hormone. It is synthesized as a glycoprotein precursor called thyroglobulin, which has 115 amino acids. Other hormones like glucocorticoids/ mineralocorticoids from the adrenal gland are synthesized and secreted in their final active form.

Pro-hormones: Some hormones are synthesized as biologically inactive or less active molecules called prohormones. Usually, they are polypeptides/ proteins. E.g., Preproinsulin→Proinsulin.

Storage:

Endocrine cells store hormones in secretory granules within their cytoplasm. For example, Thyroid hormones are stored in follicles filled with colloid particles. Cytoplasmic secretory granules store catecholamines from the adrenal medulla.

- Storage always protects the molecule from untimely inactivation.
- Steroid hormones are not stored in significant quantities.
- In response to the stimulus, they are synthesized and released immediately.

Release:

- When the target cells require free hormones, they are released immediately.

- The deficit in the bound form is replaced by the secretion of the endocrine gland. Feedback inhibition/stimulation controls hormone release.
- Protein and polypeptide hormones are released by exocytosis or pinocytosis. It involves the fusion of granules and cellular membrane, followed by secretion into the bloodstream.
- Stimulus excites the endocrine cell.
- The specific enzymes in the storage vesicle activate the hormone before release.
- Disruption of the process by certain drugs interferes with exocytosis.
- The secretory process is linked to the release of neurotransmitters.

***Transport*:**

- Some hormones are soluble and do not require transport proteins.
- The free hormone is the fraction available for binding to receptors and therefore represents the active form. Free Hormone concentration correlates best with the clinical status of either excess or deficit hormone.
- Steroid hormones are lipid-soluble. They diffuse through cell membranes.
- Specific transport proteins are found in the blood for carrying steroid hormones and thyroxine. Plasma globulins bind to thyroxine, cortisol, and sex hormones. The binding is noncovalent. Some hormones bind loosely to proteins like albumin for transport. Binding to plasma proteins protects them from inactivating systems.
- The hormones are also kept in the readily available circulatory form to the target tissues.

1.7. CONCLUSION

This chapter provided a brief introduction to biochemistry. It also discussed the foundations of biochemistry such as cellular foundations, chemical foundations, physical foundations, and so on. In this chapter, a brief introduction to the enzymes has also been discussed.

Towards the end of the chapter, it also discussed carbohydrates and glycobiology. In this chapter, amino acids, peptides, and proteins such as amino acids, peptides and proteins, and their sequences and evolution have been discussed.

REFERENCES

1. Adugna, S., Kelemu, T. and Genet, S., 2004. *Medical Biochemistry*. [ebook] USAID. Available at: <https://www.cartercenter.org/resources/pdfs/health/ephti/library/lecture_notes/health_science_students/medicalbiochemistry.pdf> [Accessed 16 June 2022].
2. Bandyopadhyay, P., Das, N. and Chattopadhyay, A., 2022. Biochemistry. *Biochemical, Immunological and Epidemiological Analysis of Parasitic Diseases*, [online] pp.245-261. Available at: <https://link.springer.com/chapter/10.1007/978-981-16-4384-2_6> [Accessed 16 June 2022].
3. GILBERT, H., 1992. *BASIC CONCEPTS IN BIOCHEMISTRY*. 2nd ed. [ebook] McGraw-Hill Companies, Inc. Available at: <https://rlmc.edu.pk/themes/images/gallery/library/books/Biochemistry/Basic%20Concepts%20in%20Biochemistry%20A%20Student's%20Survival%20Guide_IS.pdf> [Accessed 16 June 2022].
4. Louda, D., 2012. *Overview of Biomolecules*. [ebook] http://med.fau.edu. Available at: <http://med.fau.edu/students/md_m1_orientation/Overview.pdf> [Accessed 16 June 2022].
5. Nelson, D. and Cox, M., n.d. *Lehninger Principles of Biochemistry*. 4th ed. [ebook] http://aulanni.lecture.ub.ac.id. Available at: <http://aulanni.lecture.ub.ac.id/files/2012/01/15616949-Lehninger-Principles-of-Biochemistry-1-copy.pdf> [Accessed 16 June 2022].

CHAPTER 2

BIOSYNTHESIS AND IMMUNOCHEMICAL TECHNIQUES

CONTENTS

Biosynthesis is defined as a multi-step and enzyme-catalyzed process in which substrates are converted into more complex products in living organisms. In general, simple compounds are modified or converted into other compounds, or joined to form macromolecules. Initially, the chapter talks about the outline of the biosynthesis process, mainly of cell constituents. The biosynthesis of VOCs is also discussed later in the chapter along with flavonoid biosynthesis. Towards the end, immunochemical techniques such as ELISA are explained along with their working principles.

2.1. INTRODUCTION TO BIOSYNTHESIS

In living organisms, biosynthesis is a multi-step, enzyme-catalyzed process in which substrates are converted into more complex products. Simple compounds are modified, transformed into other compounds, or joined to form macromolecules during biosynthesis. This procedure frequently involves metabolic pathways.

Some of these biosynthetic pathways involve enzymes found in a single cellular organelle, while others involve enzymes found in multiple cellular organelles. The production of lipid membrane components and nucleotides are two such examples of these biosynthetic pathways. Anabolism is commonly associated with biosynthesis.

Precursor compounds, chemical energy (e.g., ATP), and catalytic enzymes that may require coenzymes are all required for biosynthesis (e.g., NADH, NADPH). These elements combine to form monomers, which serve as the building blocks for macromolecules. Proteins, which are made up of amino acid monomers linked together by peptide bonds, and DNA molecules, which are made up of nucleotides linked together by phosphodiester bonds, are two important examples of biological macromolecules.

Biosynthesis is the process by which substrates in living organisms are converted into further complicated products. The products of biosynthesis are required for cellular and metabolic processes considered necessary for survival. Biosynthesis is the anabolism branch of metabolism that produces complex proteins such as vitamins.

Biosynthetic pathways generate the vast majority of organic compounds required by microorganisms. Chemical energy and catalytic enzymes are two components used by biosynthetic pathways to promote the production of large molecules. Amino acids, purines, pyrimidines, lipids, sugars, and enzyme cofactors are examples of biosynthetic building blocks used by organisms.

Numerous mechanisms exist to ensure that biosynthetic pathways are properly controlled so that a cell produces a specific amount of a compound. The synthesis of macromolecules from specific building blocks is known as biosynthetic metabolism (also known as anabolism). The majority of these processes are multi-step or multi-enzymatic in nature.

2.2. THE BIOSYNTHESIS OF CELL CONSTITUENTS

The primary metabolic reactions that the cell uses to obtain and store energy are in the form of ATP. This metabolic energy is then used to perform a variety of tasks, such as the synthesis of macromolecules and other cell constituents. As a result, energy derived from the breakdown of organic molecules (catabolism) is used to power the synthesis of other cell components. The majority of catabolic pathways involve the oxidation of organic molecules, which results in the production of both energy (ATP) and reducing power (NADH). Biosynthetic (anabolic) pathways, on the other hand, typically involve the use of both ATP and reducing power (typically in the form of NADPH) for the production of new organic compounds.

The synthesis of carbohydrates from CO2 and H2O during the dark reactions of photosynthesis has been mentioned in the past segment as a significant biosynthetic pathway. Additional biosynthesis pathways are seen for major cellular constituents (carbohydrates, lipids, proteins, and nucleic acids).

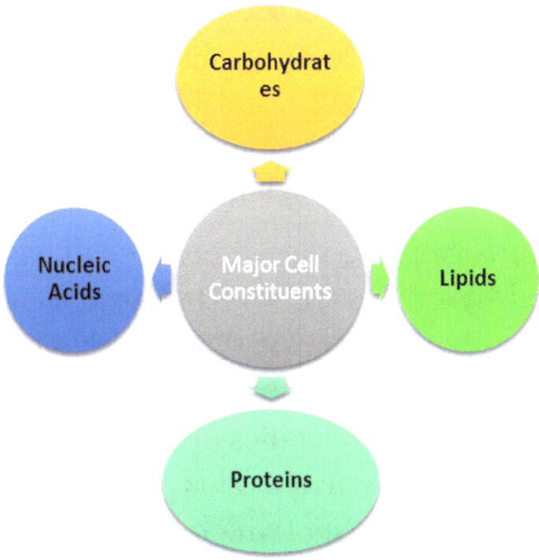

Figure 2.1. Four major constituents of a cell.

2.2.1. Carbohydrates

Glucose can be synthesized from other organic molecules in addition to being received straight from food or generated by photosynthesis. In animal cells, glucose synthesis (gluconeogenesis) typically begins with lactate (produced by anaerobic glycolysis), amino acids (derived from protein breakdown), or glycerol (produced by the breakdown of lipids).

Plants (but not animals) can also synthesize glucose from fatty acids, a process that is especially important during seed germination, once energy is stored as fats should be turned into carbohydrates to assist plant growth. Simple sugars are polymerized and stored as polysaccharides in animal and plant cells.

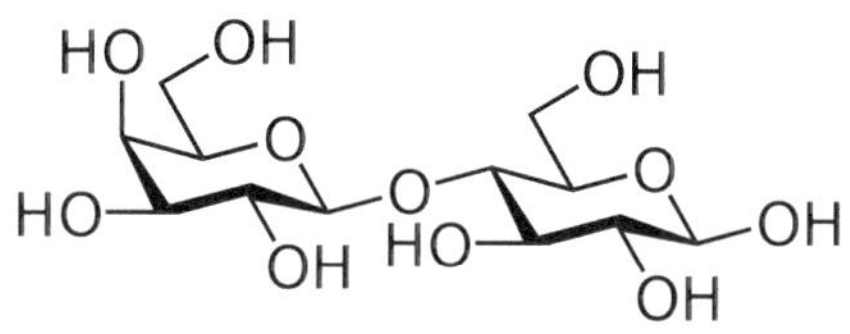

Figure 2.2. The structural formula of carbohydrates.

Source: Image by Wikimedia

Gluconeogenesis is the conversion of pyruvate to glucose, which is essentially the opposite of glycolysis. Even so, as previously discussed, the glycolytic conversion of glucose to pyruvate is an energy-yielding pathway that produces two molecules of ATP and NADH. While some glycolysis reactions are easily reversible, others would only move ahead in the position of glucose breakdown due to a significant reduction in free energy.

Throughout gluconeogenesis, these energetically favorable glycolysis reactions are bypassed by other reactions (catalyzed by various enzymes) which are paired to the expenditure of ATP and NADH in order to drive them in the direction of glucose synthesis. Overall, four molecules of ATP, two molecules of GTP, and two molecules of NADH are required to produce glucose from two molecules of pyruvate. This procedure is significantly more expensive than reversing glycolysis (which would need two molecules of ATP and two molecules of NADH), demonstrating the extra energy needed to drive the pathway in the direction of biosynthesis.

Glucose is stored in the form of polysaccharides in both plant and animal cells (starch and glycogen, respectively). Polysaccharide synthesis, like all other macromolecule synthesis, is an energy-intensive process. As previously stated, the formation of a glycosidic bond between two sugars may be represented as a dehydration event in which H2O is eliminated.

Such a reaction, however, is energetically unfavorable and hence incapable of moving ahead. As a result, the production of a glycosidic bond must be tied to an energy-yielding process, which is achieved by using nucleotide sugars as intermediates in polysaccharide synthesis. In an ATP-driven mechanism, glucose is first phosphorylated to glucose-6-phosphate, which is subsequently transformed to glucose-1-phosphate.

The reaction of glucose-1-phosphate with UTP (uridine triphosphate) produces UDP-glucose plus pyrophosphate, which is hydrolyzed to phosphate with the release of extra free energy. In an energetically advantageous reaction, UDP-glucose is an active intermediate that gives its glucose residue to a developing polysaccharide chain. Thus, chemical energy in the form of ATP and UTP promotes polysaccharide synthesis from simple sugars.

2.2.2. Lipids

Lipids are crucial energy storage molecules that makeup cell membranes. They are made from acetyl CoA, which is produced from the breakdown of carbohydrates, via a sequence of events that are similar to fatty acid oxidation in reverse. The processes leading to the synthesis of fatty acids, like those leading to their breakdown, differ from those leading to their degradation and are directed in the biosynthetic direction by being related to the expenditure of both energies in the form of ATP and reducing power in the form of NADPH.

Stepwise addition of two-carbon units produced from acetyl CoA to a developing chain produces fatty acids. Each of these two-carbon units requires one molecule of ATP and two molecules of NADPH to be added.

The 16-carbon fatty acid palmitate is the main product of fatty acid biosynthesis, which occurs in the cytoplasm of eukaryotic cells. The endoplasmic reticulum and Golgi apparatus then generate the main components of cell membranes (phospholipids, sphingomyelin, and glycolipids) from free fatty acids.

A lipid is a macro biomolecule that is soluble in nonpolar solvents in biology and biochemistry. Jab is a type of fat lipid. Non-polar liquids are generally hydrocarbons used to dissolve other naturally occurring hydrocarbon lipid molecules which do not solubilize (or do not dissolve easily) in water, such as fatty acids, waxes, sterols, fat-soluble vitamins (such as vitamins A, D, E, and K), monoglycerides, diglycerides, triglycerides, and phospholipids.

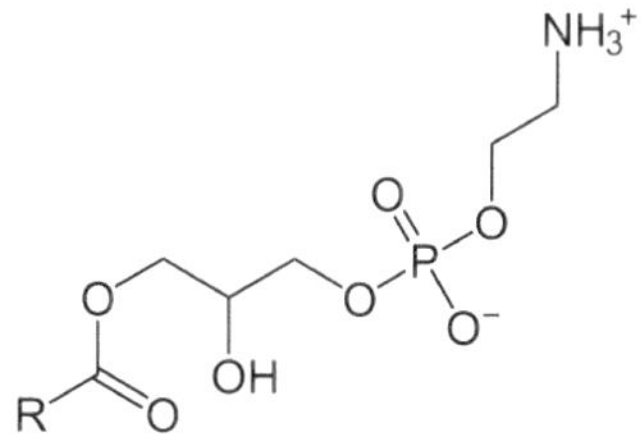

Figure 2.3. The structural formula of lipids.

Source: Image by Wikimedia

Lipids perform a variety of roles, including energy storage, signaling, and functioning as structural components of cell membranes. Lipids have uses in the cosmetic, culinary, and nanotechnology sectors.

Lipids are sometimes defined as hydrophobic or amphiphilic small molecules; the amphiphilic nature of some lipids enables them to build structures in an aqueous environment such as vesicles, multilamellar/unilamellar liposomes, or membranes. Biological lipids are made up of two types of biochemical subunits or "building blocks": ketoacyl and isoprene groups.

By using the method, lipids can be classified into eight types: fatty acids, glycerolipids, glycerophospholipids, sphingolipids, saccharolipids, and polyketides (derived from the condensation of ketoacyl subunits); and sterol lipids and prenol lipids (derived from the condensation of isoprene subunits).

2.2.3. Proteins

Proteins (as well as nucleic acids) include nitrogen in addition to carbon, hydrogen, and oxygen, as opposed to carbohydrates and lipids. Nitrogen is absorbed into organic molecules from a variety of sources in various species. Some bacteria can utilize atmospheric N2 through nitrogen fixation, which involves reducing N2 to NH3 at the price of energy in the form of ATP.

Although only a few bacteria can fix nitrogen, most bacteria, fungi, and plants can utilize nitrate (NO3-), a major element of soil, by reducing it to NH3 with electrons generated from NADH or NADPH. Lastly, ammonia (NH3) may be incorporated into organic molecules by all organisms.

The amino acids glutamatc and glutamine, which are generated from the intermediate α-ketoglutarate, are the primary sources of NH3 incorporation into organic molecules. These amino acids are subsequently used as amino group donors during the production of additional amino acids, which are likewise sourced from major metabolic pathways like the citric acid cycle.

Thus, glucose serves as the raw material for synthesis, and amino acids are produced at the expense of both energy (ATP) and reducing power (NADPH). Many microorganisms and plants are capable of producing all 20 amino acids. Humans and other animals, on the other hand, can only manufacture around half of the needed amino acids; the rest must be supplied from food.

Energy is also needed for the polymerization of amino acids into proteins. The production of the peptide bond, like the synthesis of polysaccharides, can be viewed as a dehydration process that must be directed in the direction of protein synthesis by being related to some other source of metabolic energy. This coupling occurs in polysaccharide biosynthesis by converting carbohydrates to active intermediates such as UDP-glucose. Before being employed for protein production, amino acids are similarly activated.

A key distinction between protein synthesis and polysaccharide synthesis is that amino acids are integrated into proteins in a certain order determined by a gene. The arrangement of nucleotides in a gene dictates the amino acid sequence of a protein through translation, which uses messenger RNA (mRNA) as a template for protein synthesis. In a process linked to ATP hydrolysis, each amino acid is first bonded to a particular transfer RNA (tRNA) molecule.

The aminoacyl tRNAs then bind to the ribosome-bound mRNA template, and each amino acid is added to the C-terminus of a developing peptide chain via a series of processes. During the process, two more molecules of GTP are hydrolyzed, resulting in the inclusion of each, as a result, each amino acid's incorporation into a protein is linked to the hydrolysis of one ATP and two GTP molecules.

2.2.4. Nucleic Acids

Nucleotides are nucleic acid precursors made up of phosphorylated five-carbon sugars linked to nucleic acid bases. Nucleotides can be generated from carbohydrates and amino acids received from nutrition or reused after nucleic acid degradation. The phosphorylated sugar ribose-5-phosphate, which is generated from glucose-6-phosphate, serves as the beginning point for nucleotide production. Diverse processes then lead to the synthesis of purine and pyrimidine ribonucleotides, the immediate precursors of RNA synthesis. These ribonucleotides are deoxyribonucleotides, which are the monomeric building blocks of DNA. DNA and RNA are nucleoside monophosphate polymers. In the case of other macromolecules, though, direct polymerization of nucleoside monophosphates is inefficient, and polynucleotide synthesis instead employs nucleoside triphosphates as active precursors. A nucleoside 5′-triphosphate is attached to the 3′ hydroxyl group of a developing polynucleotide chain, with pyrophosphate release and subsequent hydrolysis driving the process in the direction of polynucleotide synthesis.

Nucleic acid is a naturally occurring chemical molecule that may be dissolved to produce phosphoric acid, sugars, and an organic base combination (purines and pyrimidines). Nucleic acids are the primary information-carrying molecules in the cell, and they define the hereditary properties of all living things by controlling the process of protein synthesis.

Deoxyribonucleic acid (DNA) and ribonucleic acid (RNA) are the two primary types of nucleic acids (RNA). DNA is the master blueprint for life and is found in all free-living creatures and the majority of viruses. RNA is the genetic material of certain viruses, but it is also found in all living cells, where it is vital in activities such as protein synthesis.

2.3. BIOSYNTHESIS OF SULFUR-CONTAINING SMALL BIOMOLECULES IN PLANTS

2.3.1. Fe/S Cluster Biosynthesis in Plant Organelles

Fe/S clusters are prosthetic groups composed of acid-labile sulfur and non-heme iron which are integrated into diverse apoproteins to generate "Fe/S proteins." Mitochondria and plastids each have their unique routes for Fe/S cluster production, which include the organelle-specific L-cysteine desulfurizes NFS1 and SUFS, respectively. L-cysteine desulfurase (EC 2.8.1.7) is a pyridoxal-5'-phosphate-containing protein found in all phyla.

L-cysteine desulfurase catalyzes the process which eliminates a sulfur atom from the L-cysteine substrate to create L-alanine. Throughout this catalytic process, L-cysteine desulfurase transiently binds the sulfur generated from L-cysteine as sulfane sulfur in the form of persulfide, whereupon the sulfur bound to the enzyme is transferred to a Fe/S biosynthetic "scaffold protein" or its complex. Mitochondria and plastids each have their own scaffold proteins that produce an instability nascent Fe/S cluster.

The nascent unstable Fe/S clusters are then formed and matured with the help of additional Fe/S carrier proteins and chaperones before being integrated into the target proteins' apo-forms. Because many Fe/S proteins play essential roles in the electron transfer systems of respiration and photosynthesis, both the mitochondrial nitrogen fixation protein for sulfur transfer (NFS) pathway and the plastidial sulfur utilization factor (SUF) pathway are required for these energy-producing organelles.

NFS1 acts as a sulfur donor in mitochondria, while ISD11 is assumed to stabilize NFS1. Sulfur that binds to NFS1 transiently is absorbed into the mitochondrial scaffold ISU proteins, forming an unstable Fe/S cluster. Three ISU proteins have been found in Arabidopsis mitochondria, with ISU1 appearing to be a key scaffold protein due to the low expression levels of the other two ISU proteins (ISU2 and ISU3). ISA1 and ISCA4 have also been linked to the production of mitochondrial Fe/S clusters, potentially as carrier proteins.

The mitochondrial Fe/S cluster biosynthesis is also suggested to be aided by the molecular chaperones HSCA1, HSCA2, and HSCB. Several additional proteins, including to generate [4Fe-4S]-type Fe/S clusters, NFU5/NFU-I and NFU4/NFU-III are needed. According to a recent study, these NFU proteins are reduced through the TRX -mediated redox system rather than the GRX -mediated system.

IBA57 and INDH are proteins that help the Fe/S cluster in mitochondrial respiratory chain complex I mature. It has also been observed that the mitochondrial monothiol glutathione oxidoreductase GRXS15 is essential for the production of Fe/S clusters.

While several protein components essential for mitochondrial Fe/S production have been found and described, their specific biochemical activities and functional cooperation remain unknown. Furthermore, the molecular machinery necessary for the production and inclusion of diverse forms of Fe/S clusters into apoproteins is yet unknown.

Based on the similarities to the specific bacterial SUF system, it is postulated that SUFS give sulfane sulfur to a scaffold complex composed of SUFB, SUFC, and SUFD and contributes to the first biosynthetic process in Arabidopsis plastidial Fe/S cluster production. SUFE1 interacts with SUFS and is hypothesized to be involved in cysteine sulfur mobilization.

2.4. BIOSYNTHESIS OF VOLATILE PLANT SECONDARY METABOLITES AND ITS INTERCONNECTION WITH PRIMARY METABOLISM

Volatile organic compounds (VOCs), lipophilic liquids with low molecular weight and high vapor pressure at room temperature, make up a sizable subset of plant natural products. These molecules' physical features allow them to readily pass cellular membranes and be discharged into the surrounding environment. Over the years, over 1700 VOCs from 90 various plant groups, including angiosperms and gymnosperms, have been found (Knudsen et al., 2006).

VOC biosynthesis is reliant on the availability of carbon, nitrogen, and sulfur, as well as energy from primary metabolism. As a result of the high degree of connectedness between primary and secondary metabolism, the availability of these building blocks has a significant influence on the concentration of any secondary metabolite, including VOCs.

Only a few basic metabolic processes are involved in the biosynthesis of the diverse range of VOCs. All VOCs are classified into numerous groups based on their biosynthetic origin, including terpenoids, phenylpropanoids/ benzenoids, fatty acid derivatives, and amino acid derivatives, as well as a few species-/genus-specific molecules that are not included in those core categories.

2.4.1. Biosynthesis of Terpenoids

Terpenoids are the most abundant and diversified family of secondary metabolites, with several volatile compounds produced from mainly two five-carbon precursors, isopentenyl diphosphate (IPP) and its allylic isomer, dimethylallyl diphosphate (DMAPP). These C5-isoprene building blocks are formed in plants by two distinct, compartmentally separated pathways:

mevalonic acid (MVA) and methylerythritol phosphate (MEP). The MVA route produces volatile sesquiterpenes (C15), whereas the MEP pathway produces volatile hemiterpenes (C5), monoterpenes (C10), and diterpenes (C20).

Based on the experimental data and projections of their subcellular localization, the MEP route is thought to be entirely plastidic, as a complete set of the relevant enzymes occurs only in plastids. The subcellular location of the MVA route, on the other hand, is less known. This pathway was formerly thought to be cytosolic; however, fresh data reveals that the MVA pathway is dispersed throughout the cytosol, endoplasmic reticulum, and peroxisomes.

The MVA pathway is composed of six enzymatic processes that begin with the stepwise condensation of three molecules of acetyl-CoA to 3-hydroxy-3-methylglutaryl-CoA, which is then reduced to MVA by two successive phosphorylation and a decarboxylation/elimination step, with IPP as the ultimate result. Because acetyl-CoA cannot easily traverse membranes, it is unclear which subcellular pool is employed for terpenoid biosynthesis.

Pools exist in chloroplasts, peroxisomes, mitochondria, cytosol, and nucleus. The Arabidopsis genome has two genes producing acetoacetyl-CoA thiolase (AACT), one of which (AACT2) catalyzes the first step in the MVA pathway and, according to proteome analysis, is located in peroxisomes.

The MEP route consists of seven enzymatic steps that started with the fusion of d-glyceraldehyde 3-phosphate (GAP) and pyruvate (PYR) to create 1-deoxy-d-xylulose 5-phosphate, which is subsequently isomerized/reduced to form the pathway's distinctive intermediate, MEP. Converting MEP to IPP and DMAPP requires five stages in succession. PYR and GAP are supplied by primary metabolism to the MEP system, with the latter obtained from both glycolysis and the pentose phosphate pathway (PPP).

To date, the origin of PYR in chloroplasts is unknown, as plastids have poor activity of the essential glycolytic enzymes, phosphoglycerate mutase and enolase, and may be unable to meet the high PYR requirement for isoprenoid production. Indeed, in *Arabidopsis thaliana*, plastid-localized IPP production was disrupted in *Arabidopsis thaliana* mutants missing a PYR transporter, which provides cytosolic PYR to the MEP route.

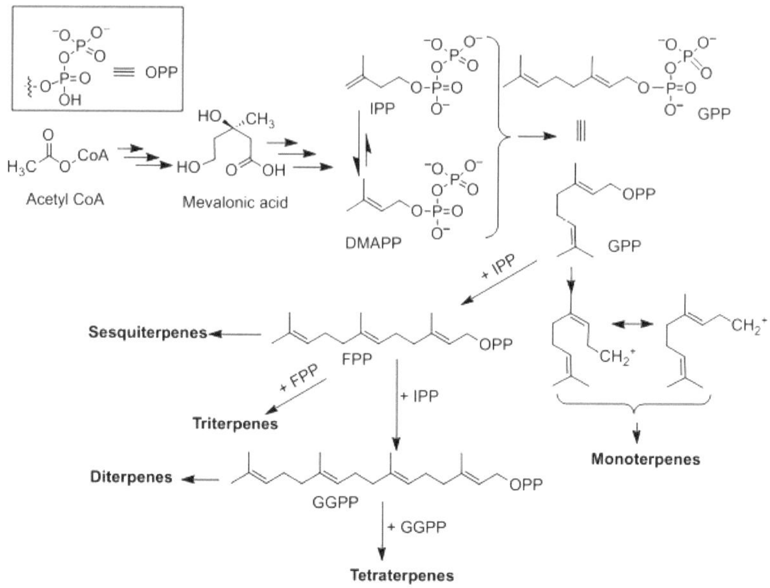

Figure 2.4. Terpene biosynthesis.

Source: Image by pixabay

2.4.2. Biosynthesis of Phenylpropanoid/Benzenoid Compounds

The aromatic amino acid phenylalanine (Phe) is the source of the second biggest family of plant VOCs, phenylpropanoid and benzenoid compounds. Phe is linked to central carbon metabolism via seven enzyme processes of the shikimate pathway and three of the arogenate pathway. The shikimate pathway's immediate precursors, phosphoenolpyruvate (PEP) and d-erythrose 4-phosphate (E4P), are derived from glycolysis and the PPP, respectively.

Because the same processes produce precursors for the MEP system, the latter must compete for carbon allocation with the shikimate/phenylpropanoid pathway, especially as Phe receives 30% of photosynthetically fixed carbon, mostly to create lignin. 3-deoxy-D-arabino-heptulosonate 7-phosphate synthase (DAHP synthase) is the first gene in the shikimate pathway and is important in modulating carbon input into the process. The molecular pathways, But, the molecular processes behind this control in plants are mainly unexplored.

Although Phe production occurs in plastids, it is converted to volatile chemicals outside of this organelle. The very first dedicated step in the biosynthesis of the majority of phenylpropanoids/benzenoids is catalyzed by PAL, a well-known and extensively distributed enzyme that deaminates Phe to trans-cinnamic acid (CA).

The formation of benzenoids (C6-C1) from cinnamic acid needs a two-carbon shortening of the propyl side chain and has been demonstrated to occur via a β-oxidative mechanism, a non-oxidative pathway, or a combination of these methods. The β-oxidative pathway, which appears to be similar to that used in the catabolism of fatty acids and some branched-chain amino acids, was only recently fully described.

This route starts with the activation of CA to its CoA thioester, which then proceeds through hydration, oxidation, and cleavage to generate benzoyl-CoA. The -oxidative pathway's peroxisomal location raises the topic of benzoyl-CoA export to the cytoplasm for benzyl benzoate and phenylethyl benzoate production.

2.4.3. Biosynthesis of Volatile Fatty Acid Derivatives

The other kind of plant VOC is fatty acid derivatives, which include 1-hexanal, cis-3-hexenol, nonanal, and methyl jasmonate and are derived from C18 unsaturated fatty acids, linoleic or linolenic. These fatty acids are biosynthesized using a plastidic pool of acetyl-CoA produced from PYR, the end result of glycolysis.

Unsaturated fatty acids undergo stereospecific oxygenation after entering the 'lipoxygenase (LOX) route,' giving 9-hydroperoxy and 13-hydroperoxy intermediates, that are then metabolized via the two branches of the LOX pathway, releasing volatile chemicals. The allene oxide synthase branch generates jasmonic acid by using solely the 13-hydroperoxy intermediate (JA), JA carboxyl methyl transferase converts this to methyl jasmonate.

The hydroperoxide lyase branch, on the other hand, transforms both types of hydroperoxide fatty acid derivatives into C6 and C9 aldehydes, which are frequently reduced to alcohols by alcohol dehydrogenases before being further converted to esters. These saturated and unsaturated C6/C9 aldehydes and alcohols, also known as green leaf volatiles, are typically generated in green organs of plants in reaction to injury, but they also supply vegetables and fruits with their distinctive 'fresh green' fragrance.

2.4.4. Biosynthesis of Volatiles Derived From Branched-Chain Amino Acids

Many volatile chemicals, particularly those seen in flower smells and fruit odors, are produced from amino acids or intermediates in their formation and include nitrogen and sulfur. The manufacture of these amino acid-derived volatiles in plants is thought to happen in a manner similar to those observed in bacteria or yeast, in which these pathways have received greater attention.

As in microorganisms, amino acids undergo initial deamination or transamination catalyzed by aminotransferases, resulting in the creation of the matching -ketoacid. These -ketoacids can then be decarboxylated, proceeded by reductions, oxidations, and/or esterification to generate aldehydes, acids, alcohols, and esters. Amino acids can also be acyl-CoA precursors. Amino acids also can be employed as a precursor for acyl-CoAs, which are utilized in alcohol esterification processes mediated by alcohol acyltransferases (AATs).

2.5. FLAVONOID BIOSYNTHESIS

From a genetic standpoint, much work has been put into deciphering the biosynthesis processes of flavonoids. Flavonoid synthesis mutants were isolated from a variety of plant species. As the first significant experimental models in this system, maize (*Zea mays*), snapdragon (*Antirrhinum majus*), and petunia (*Petunia hybrida*) were created, leading to the discovery of several structural and regulatory flavonoid genes. Recently, Arabidopsis (*Arabidopsis thaliana*) has aided in the investigation of the flavonoid pathway's regulation and subcellular localization.

The use of Arabidopsis for flavonoid biosynthesis is fascinating since single copy genes encode all enzymes of the core flavonoid metabolism, with the exception of flavanol synthase (FLS), which is encoded by six genes but only two (FLS1 and FLS3) have shown activity. Mutations that erase or diminish seed coat pigmentation were used to identify genetic loci for both structural and regulatory genes; hence, the loci were dubbed transparent testa or tt mutants.

As a result, the majority of structural genes, as well as a few regulatory genes, have been linked to particular mutant loci in Arabidopsis. This species doesn't quite appear to utilize flavonoids in the same manner that the other species do (for example, in defense or for male fertility); yet, these mutants are useful in defining functions for these substances in critical processes such as UV protection and auxin transport control.

The phenylpropanoid route converts phenylalanine into 4-coumaroyl-CoA, which then enters the flavonoid biosynthesis pathway. Chalcone synthase, the first flavonoid pathway enzyme, generates the chalcone scaffolds from which all flavonoids are derived. Even though the central pathway for flavonoid biosynthesis is preserved in plants, depending on the species, a group of enzymes such as isomerases, reductases, hydroxylases, and several Fe2+/2-oxoglutarate-dependent dioxygenases modify the basic flavonoid skeleton, resulting in the various flavonoid subclasses. Finally, transferases change the flavonoid backbone with sugars, methyl groups, and/or acyl moieties, regulating the physiological activity of the resultant flavonoid by changing its solubility, reactivity, and interaction with cellular targets.

Increasing evidence suggests suggesting sequential enzymes of the phenylpropanoid and flavonoid biosynthesis are structured into macromolecular complexes which can be connected to endomembranes. Metabolic channeling in plant secondary metabolism allows plants to successfully produce certain natural compounds while avoiding metabolic interference.

The presence of cytochrome P450 monooxygenases (P450s)-related metabolons has been demonstrated: direct and indirect experimental evidence characterize P450 enzymes in the phenylpropanoid, flavonoid, cyanogenic glucoside, and other biosynthetic pathways. Transgenic tobacco plants expressing epitope-tagged versions of two phenylalanine ammonia lyase isoforms (PAL1 and PAL2) and cinnamate-4-hydroxylase have been used to provide additional evidence for the channeling of intermediates between specific isoforms of phenylalanine ammonia lyase and cinnamate-4-hydroxylase. Furthermore, yeast-two hybrid tests have suggested the existence of a multienzyme complex for the anthocyanin pathway in rice.

The majority of flavonoid synthesizing enzymes are recovered in soluble cell fractions; immunolocalization experiments indicate that they will be loosely bound to the endoplasmic reticulum (ER), possibly in a multi-enzyme complex, while the pigments themselves accumulate in the vacuole (i.e., anthocyanins and proanthocyanidins) or the cell wall. Flavanol synthase 1, and also chalcone synthase and chalcone isomerase, have recently been discovered in Arabidopsis nuclei.

Antirrhinum majus aureusidin synthase, the enzyme that catalyzes aurone biosynthesis from chalcones, was found in the vacuole, whereas chalcone 4′-O-glucosyltransferase was found in the cytoplasm, indicating that chalcones 4-O-glucosides are transported to the vacuole and converted to aurone 6-O-glucosides there. Furthermore, a flavonoid-3′-hydroxylase was recently discovered in the tonoplast of the soybean immature seed coat's hilum region.

The mechanisms of anthocyanin transfer from the ER to the vacuole storage sites have been postulated by two models: ligandin transport and vesicular transport. The ligandin transport paradigm is based on genetic evidence that glutathione transferase (GST)-like proteins are essential for pigment vacuolar sequestration in maize, petunia, and Arabidopsis (AtTT19).

The vacuolar sequestration of anthocyanins in maize requires a multidrug resistance related protein-type (MRP) transporter on the tonoplast membrane, whose expression is co-regulated with structural anthocyanin genes. MRP proteins are also known as glutathione S-X (GS-X) pumps because they transport a range of glutathione conjugates.). Furthermore, since no anthocyanin–glutathione conjugate(s) were discovered, it is speculated that these GSTs may convey their flavonoid substrates straight to the transporter, functioning as a carrier protein or ligandin.

This theory is reinforced by the fact that Arabidopsis' GST (TT19), which is found in both the cytoplasm and the tonoplast, may attach to glycosylated anthocyanins and aglycones but does not conjugate them with glutathione. The hypothesized vesicle-mediated transport concept is based on observations that anthocyanins and other flavonoids accumulate in the cytoplasm in discrete vesicle-like structures (anthocyanoplasts) before being imported into the vacuole via an autophagic process.

Nonetheless, grape vesicle-mediated anthocyanin transport requires a GST as well as two multidrug and hazardous chemical extrusion-type transporters (anthoMATEs). Thus, our data suggest the presence of both transport systems, in which the engagement of GSTs and transporters would be cell and/or flavonoid-specific.

2.5.1 Regulation of Flavonoid Biosynthesis

The interplay of several transcription factor families regulates the flavonoid biosynthesis genes. R2R3 MYB transcription factors, basic helix-loop-helix (bHLH), and WD40 proteins differently control genes implicated in the anthocyanin pathway in monocot and dicot species. Therefore, the activation and spatial and temporal expression of anthocyanin biosynthesis structural genes are determined by combinations of the R2R3-MYB, bHLH, and WD40 transcription factors and their interactions (MYB-bHLH-WD40 complex). MYB-bHLH-WD40 complex control of anthocyanin production in reproductive and other organs has been reviewed.

Anthocyanin regulation differs markedly between monocot and dicot species such as Arabidopsis and maize. TTG1, a WD40 transcription factor, various bHLH (TT8, GL3, and EGL3), and MYB transcription factors (PAP1 and PAP2) interact in Arabidopsis to trigger anthocyanin production in vegetative tissues. MYB and bHLH proteins in maize are encoded by two multigene families (PL/C1 and B/R, respectively), with each member having a tissue- and developmental-specific pattern, whereas WD40 protein PAC1 is needed by both B1 and R1 proteins for full activation of anthocyanin biosynthetic genes in seeds and roots.

Functional Arabidopsis that works TTG1 is essential for anthocyanin accumulation throughout the development of roots and trichomes, and maize PAC1 can complement Arabidopsis ttg1 mutants; however, maize pac1 mutants only display a decrease in anthocyanin pigmentation in certain tissues. Furthermore, the control of flavanol production differs significantly between the two species. Three R2R3-MYB proteins, MYB12, MYB11, and MYB111 (PFG1-3), which have diverse spatial expression patterns in Arabidopsis, control AtFLS1 expression in a tissue- and developmental-specific way.

ZmFLS1/2, on the other hand, is controlled by both P1 (R2R3-MYB) and the anthocyanin C1/PL1 and R/B regulators. Despite the fact that flavanols are required for pollen germination and conditional male fertility in maize, plants lacking the anthocyanin regulators P1 and R/B+C1/PL1 are fertile. Furthermore, PFG1-3-independent flavanol accumulation occurs in Arabidopsis pollen and siliques/seeds, indicating that other regulators, still to be found, also are engaged in the control of FLS expression, and therefore flavanol accumulation, in both species.

Furthermore, the structural and functional development of the MYB and bHLH plant families has been thoroughly studied. Interestingly, the discovery of a C1-like (MBF1) regulator in the gymnosperm *Picea mariana* (black spruce) supports the theory that anthocyanin pathway control by a C1-like class of R2R3 MYB protein predates the evolutionary divergence of angiosperms and gymnosperms. The discovery of bHLH and MYB proteins in mosses lends credence to the theory that the bHLH–MYB complex originated earlier during land plant evolution.).

Various R2R3 MYB transcription elements were discovered in maize, Antirrhinum, Petunia, and Arabidopsis. These transcription factors are involved in the control of the flavonoid biosynthesis pathway. The growing availability of plant genomes has facilitated the identification and separation of a large number of MYB genes involved in the regulation of flavonoid biosynthesis from a variety of non-model plant species including grapevine (*Vitis vinifera*), strawberry (*Fragaria x ananassa*), apple (*Malus domestica*), cauliflower *(Brassica oleracea var botrytis*), potato (*Solanum tuberosum* L.), bayberry (*Myrica rub*) and kale (*Brassica oleracea var. acephala f. tricolor*). Furthermore, several of these regulators have been functionally defined by transient studies and persistent expression in heterologous vegetal hosts.

2.6. IMMUNOCHEMICAL TECHNIQUES

Every vertebrate has an advanced immune system. The immune system becomes more advanced as the organism becomes more complicated. Mammalian immune systems have developed over a million years, providing an extraordinary defense mechanism capable of reacting to infective threats that emerge in the body. Immunity in our bodies is regulated or controlled by particular cells derived from bone marrow stem cells. B and T lymphocytes are the most significant cell types because they can fight germs and viruses. B-cells produce antibodies in response to the introduction of a foreign material into the body.

An antigen is a foreign material that can elicit an immunological response that results in the creation of antibodies. They are the sites where antibodies bind. As a result, antibodies are antigen-specific (bind to the antigen that initiated its production). This section is concerned with the identification and diagnostic chemical procedures used to assess an antibody's reaction to a specific antigen. Immunochemical approaches are based on antibodies' selective, reversible, and non-covalent binding to antigens. These techniques are used to detect or quantify antigens or antibodies.

Immunochemistry is a branch of immunology that is well sophisticated. This is concerned with the chemical components and chemistry (chemical processes) of immunological phenomena, namely an antibody and an antigen. Immunochemical approaches make use of an antibody's highly specific affinity for its antigen. It identifies the presence of a certain protein or antigen in tissues or cells. Immunochemical procedures are the methods used for immunochemical analysis; they are extremely significant in diagnostic and clinical contexts, as even normal cells with numerous proteins are changed in pathological states (in cancer).

2.6.1. Radioimmunoassay

Radioimmunoassay (RIA) seems to be a very sensitive *in vitro* assay method that uses antibodies to assess antigen concentrations (for example, hormone levels in the blood). As such, it is the inverse of a radio binding test, that measures an antibody using equivalent antigens. Despite the fact that the RIA approach is exceedingly sensitive and specific, requiring specialist equipment, it is still the least expensive means of doing such tests.

Because radioactive compounds are utilized, specific precautions and authorization are required. It has now been overtaken by the ELISA approach, which measures the antigen-antibody interaction using colorimetric signals rather than a radioactive signal. However, due to their robustness, consistency, and cheap cost per test, RIA approaches are regaining popularity. It is often easier to carry out than a bioassay.

A radioimmunoassay is the RAST test (radioallergosorbent test). This is used to identify the allergen that is causing an allergy.

Method of Radioimmunoassay

To conduct a radioimmunoassay, a measured amount of an antigen is rendered radioactive, often by labeling it with gamma-radioactive iodine isotopes linked to tyrosine. This radiolabeled antigen is then combined with a predetermined quantity of antigen-antibody, and the two chemically bond to one another. Then a patient serum sample having an unknown amount of that same antigen is introduced. As a result, the unlabeled (or "cold") antigen from the serum competes for antibody binding sites with the radiolabeled antigen ("hot").

"Colder" antigen attaches to the antibody as its concentration increases. Displacement of radiolabeled variant and decreasing the antibody-bound radiolabeled antigen/free radiolabeled antigen ratio. The bound antigens are isolated from the free antigens, and the radioactivity of the free antigen left in the supernatant is quantified with a gamma counter. A binding curve may then be built using established standards, allowing the quantity of antigen in the patient's serum to be calculated.

2.6.2. ELISA

The enzyme-linked immunosorbent assay (ELISA) has been utilized as a diagnostic tool in medicine and plant pathology, as well as a quality-control check in a variety of sectors. The following steps are taken to detect (and quantify) the presence of the antigen in the sample: The sample's antigens are bonded to a surface. Then, a second particular antibody is given to the surface to attach to the antigen. This antibody is coupled to an enzyme, and then a material containing the enzyme's substrate is added in the last step. The ensuing reaction generates a visible signal, most typically a change in the color of the substrate.

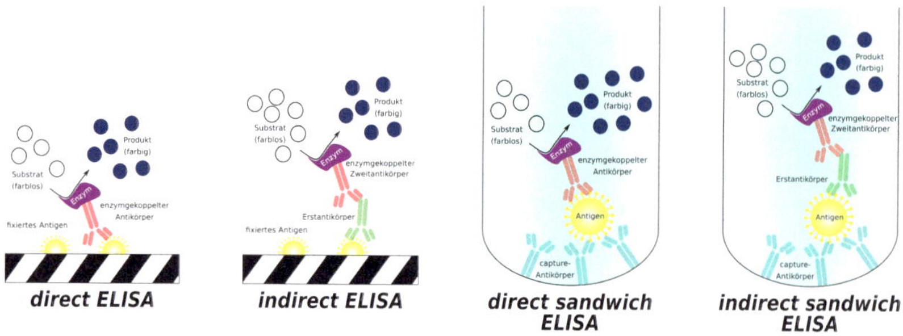

Figure 2.5. Types of ELISA.

Source: Image by Wikimedia

An ELISA requires at least one antibody that recognizes a specific antigen. A sample containing an unknown quantity of antigen is immobilized on a solid substrate (often a polystyrene microtiter plate) either non-specifically (through adsorption to the surface) or specifically (by immobilization to the surface) (via capture by another antibody specific to the same antigen, in a "sandwich" ELISA). The detecting antibody is added after the antigen has been immobilized, generating a combination with the antigen.

The detection antibody can be covalently coupled to an enzyme or identified by a secondary antibody that has been bioconjugated to an enzyme. During steps, the plate is normally cleaned with a mild soap attempt to eliminate any unbound proteins or antibodies. Following the final washout step, the plate is made by adding an enzyme substrate to provide a visual signal indicating the amount of analyte in the sample.

Notably, ELISA may perform different types of ligand binding assays, in addition, to purely "immuno" assays, yet the term retained the original "immuno" because of the method's widespread use and history of development. The approach simply requires any ligating reagent that can be immobilized on the solid phase, as well as a detection reagent that will bind selectively and employ an enzyme to give a quantifiable signal.

Between washes, only the ligand and its particular binding equivalents remain precisely bound or "immunosorbed" to the solid phase through antigen-antibody interactions, while nonspecific or unbound components are washed away. Unlike other spectrophotometric wet lab assay forms in which the same response well (e.g., a cuvette) may be repeated after cleaning, because the reaction products are immunosorbed on the solid form that is a component of the plate, the ELISA plates cannot be easily reused.

Principle of ELISA

As a "wet lab" analytic biochemistry assay, ELISA involves detecting an "analyte" (i.e. the specific substance whose existence is being quantitatively or qualitatively analyzed) in a liquid sample using a technique that persists to use liquid reagents during the "analysis" (i.e. controlled sequence of biochemical reactions that will generate a signal that can be easily measured as well as interpreted as a measurement of the amount of analyte. In contrast to "dry tests," which can employ dry strips, the final detection phase in "dry" analysis involves reading a dried strip using techniques including reflectometry, but does not require a response containment vessel to avoid spillage or sample mixing.

As a heterogenous assay, ELISA isolates some components of the analytical reaction mixture by adsorbing specific components onto a physically immobilized solid phase. In ELISA, a liquid sample is applied to a stationary solid phase with special binding properties, accompanied by numerous liquid reagents which are successively added, incubated, and cleaned, preceded by certain optical changes (e.g.,color development by an enzymatic reaction product) in the last liquid in the well through which the analyte amount is assessed. The qualitative "reading" is generally based on spectrophotometry, which includes quantifying the transmission of a given wavelength of light through the liquid (Also, with the multiple-well plate configuration, the translucent bottom of the well).).

2.6.3. Sandwich ELISA

Sandwich ELISA is a handy format for bigger target molecules with multiple epitopes. The firm support is coated with one antibody generated against one epitope of the drug; the targeted molecule is then added, following the second antibody raised against another epitope of the molecule. A second antibody may well be labeled, or a third labeled antibody may be employed for detection. Because OTA has a low molecular mass and fits into the binding pocket of the antibody, there is no additional moiety exposed for the binding of the second antibody.

A sandwich ELISA method, therefore, has been developed for detecting various molds (Aspergillus, Penicillium, Fusarium) concurrently by utilizing monoclonal antibodies against such species' extracellular polysaccharides. Punyatong et al. (2003), on the other hand, developed a sandwich ELISA employing two monoclonal antibodies manufactured in-house using OTA-HSA as the immunogen. The reported ELISA, which included one MAb coated on a microtiter plate and one HRP-labeled MAb, demonstrated 50% binding at 35 pg/assay, which is highly sensitive. There is no additional validation or application described. The assay was able to detect B. cereus when the samples were prepared in meat with various pathogens. The newly developed analytical method provides a rapid method to sensitively detect *B. cereus* in food specimens.

2.7. PRINCIPLES OF IMMUNOCHEMICAL TECHNIQUES USED IN CLINICAL LABORATORIES

Immunochemistry provides simple, quick, robust but sensitive procedures that are easily automated and relevant to regular clinical laboratory tests. Immunochemical approaches often do not need lengthy and damaging sample preparation or costly apparatus. In reality, the majority of approaches rely on basic photo-, Fluoro-, or luminometric detection.

Immunochemical approaches have quickly surpassed chromatographic techniques in clinical diagnostics, allowing for the rapid identification of antibodies linked to specific medical disorders, disease biomarkers, hormones, and medications. The most common assays used during clinical immunochemistry are quantitative or qualitative formats utilizing enzyme-linked immunosorbent assays (ELISAs), and immunochromatography in the form of lateral-flow devices like dip-sticks and test strips, or Western Blot assays used to interpret data from gel electrophoresis. Similarly, immunohistochemistry, among the most important diagnostic techniques in today's medical laboratories, is founded on antigen-antibody binding principles.

2.7.1. Antigen-Antibody Binding

All immunochemical techniques rely on an extremely specific and sensitive interaction between an antigen and an antibody. The antigen is a substance that stimulates the creation of antibodies [that is, proteins from the class of immunoglobulins (MW around 150 kDa) generated in the immune system of any animal or human as a result of a defense reaction (immunity) to this foreign material].

Antibodies are a diverse group of glycoproteins with similar structural and functional features. They are distinguished by their capacity to attach to both antigens and specialized immune system cells or proteins. Antibodies are frequently shown structurally as Y-shaped structures with four polypeptides—two identical polypeptide units known as heavy chains and two additional polypeptide units known as light chains.

Immunoglobulins are classified into five types depending on the number of Y-like units and the kind of heavy-chain polypeptide, and they contain: IgG, IgM, IgA, IgE, and IgD. The most common serum antibody is IgG, which has three protein domains, two of these are similar (Fab fragments) and create the arms of the Y. Each Fab region has a binding site for an antigen. The third domain (complement-binding Fc fragment) serves as the Y's foundation and is essential for immune system activity and control.

An epitope or immuno-determinant region is the area of an antigen that reacts with an antibody. The attachment of an antibody to the antigen is based on reversible, noncovalent interactions, and the complexes are in equilibrium with the unbound components. An antibody's binding pocket may hold 6 to 10 amino acids. Small variations in antigen structure (such as a single amino acid) can influence the intensity of the antibody-antigen interaction.

The strength of the binding is assessed by affinity, which is commonly quantified in terms of the concentration of an antibody-antigen complex measured at equilibrium. It normally varies from micro (10-6) to pico (10-12) molar. High-affinity antibodies may bind more antigens in less time and create more stable complexes than low-affinity antibodies. As a result, high-affinity antibodies are typically selected in immunochemical procedures.

Avidity is another metric used to define the antibody-antigen binding process. This is a measurement of the complex's overall stability, governed by the antibody's affinity for the epitope, the number of binding sites per antibody molecule, and the geometric arrangement of the interacting components. Avidity defines all elements involved in the binding process and influences the effectiveness of all immunochemical procedures. Antibodies are often very selective for the antigen.

Nevertheless, some antibodies exhibit cross-reactivity to comparable epitopes on different molecules. As a result, the immunochemical technique is less specific, but more useful in cases when the target is a class of structurally similar compounds. The quality of the antibodies employed in the procedure determines the specificity and sensitivity of the assay.

Polyclonal antibodies are recovered from the serum of vaccinated animals, generally, rabbits and the serum contain a variety of antibodies with varying specificities and affinities. Because of the vaccination of mice followed by hybridoma technologies, monoclonal antibodies are more homogeneous in terms of specificity and affinity.

2.7.2. Precipitin Test

The precipitin test is an illustration of an antigen-antibody reaction-based clinical diagnostic. Antibodies may precipitate antigens by multivalent binding, which occurs when two Fab fragments of a single antibody bind to two antigens at the same time. A solution containing an antigen: antibody complex matrix will result in the production of a visible precipitate. When a soluble antigen and an antibody meet at an appropriate concentration, a visible precipitate occurs.

The double immunodiffusion also called Ouchterlony gel diffusion occurs in an agarose gel with neighboring wells loaded with an unknown soluble antigen and a known antibody solution. If an antigen suitable for the identified antibody is present, the two components combine to form a visual precipitin band in the gel. Electrical current is used in counter – immuno - electrophoresis (CIE) to accelerate the migration of soluble antigens and antibodies.

Many microbial antigens contain a net negative charge that migrates towards the positively charged electrode at basic pH. Antibody molecules, on the other hand, are only slightly negatively charged or neutral in alkaline circumstances, thus they do not move in the electrical field but are transported toward the cathode by the buffer ions—a process known as electroendosmosis.

When the target antigen and antibody come into contact, they generate a detectable precipitin band. The electrophoretically-enhanced mechanism is substantially quicker than passive diffusion, taking less than an hour to complete. CIE was frequently employed in clinical laboratories in the 1970s, however, it has been superseded by quicker and more accurate immunochemical techniques for antigen detection.

2.7.3. Immunoturbidimetric Assays

Microparticle immunoturbidimetric tests could be used to quantify medicines or biomarkers in bodily fluids such as serum, plasma, or urine. These tests are based on an agglutination response caused by antigen-antibody binding. For analysis of drugs, the test frequently comprises a competition between drugs in the sample and drug-coated onto a microparticle. In the absence of a free drug, the drug-coated microparticle is quickly agglutinated in the presence of antibodies.

When light is focused on the sample mixture, the photometric absorption change is proportional to the speed of agglutination of the microparticles. If there is free medicine in the sample, the agglutination process is somewhat blocked, delaying the rate of absorbance change. A concentration-dependent classic agglutination curve may be generated, with the minimum drug concentration yielding the highest percentage of agglutination and the highest drug concentration yielding the lowest rate of agglutination. Serum ferritin, for example, may be assessed utilizing competition between protein in the sample and antigen for a particular antibody coated onto a microparticle.

2.8. CONCLUSION

Numerous mechanisms exist to ensure that biosynthetic pathways are properly controlled so that a cell produces a specific amount of a compound. The production of lipid membrane components and nucleotides are two such examples of these biosynthetic pathways. Amino acids, purines, pyrimidines, lipids, sugars, and enzyme cofactors are examples of biosynthetic building blocks used by organisms.

Chemical energy and catalytic enzymes are two components used by biosynthetic pathways to promote the production of large molecules. Biosynthetic pathways generate the vast majority of organic compounds required by microorganisms. These elements combine to form monomers, which serve as the building blocks for macromolecules. Some of these biosynthetic pathways involve enzymes found in a single cellular organelle, while others involve enzymes found in multiple cellular organelles.

The synthesis of macromolecules from specific building blocks is known as biosynthetic metabolism (also known as anabolism). Simple compounds are modified, transformed into other compounds, or joined to form macromolecules during biosynthesis. The products of biosynthesis are required for cellular and metabolic processes considered necessary for survival.

Precursor compounds, chemical energy (e.g., ATP), and catalytic enzymes that may require coenzymes are all required for biosynthesis (e.g., NADH, NADPH). In living organisms, biosynthesis is a multi-step, enzyme-catalyzed process in which substrates are converted into more complex products.

Biosynthesis is the anabolism branch of metabolism that produces complex proteins such as vitamins. Biosynthesis is the process by which substrates in living organisms are converted into further complicated products. Anabolism is commonly associated with biosynthesis. This procedure frequently involves metabolic pathways.

REFERENCES

1. 2022. Biosynthesis and Chemical Properties of Natural Substances in Plants. [ebook] Available at: <https://application.wiley-vch.de/books/sample/3527332308_c01.pdf> [Accessed 2 July 2022].
2. 2022. IMMUNOCHEMICAL TECHNIQUES. [ebook] Available at: <https://nios.ac.in/media/documents/dmlt/Biochemistry/Lesson-24.pdf> [Accessed 2 July 2022].
3. Cooper, G., 2022. The Biosynthesis of Cell Constituents. [online] Ncbi.nlm.nih.gov. Available at: <https://www.ncbi.nlm.nih.gov/books/NBK9956/> [Accessed 2 July 2022].
4. Dudareva, N., Klempien, A., Muhlemann, J. and Kaplan, I., 2013. Biosynthesis, function and metabolic engineering of plant volatile organic compounds. New Phytologist, [online] 198(1), pp.16-32. Available at: <https://nph.onlinelibrary.wiley.com/doi/10.1111/nph.12145> [Accessed 2 July 2022].
5. Falcone Ferreyra, M., Rius, S. and Casati, P., 2012. Flavonoids: biosynthesis, biological functions, and biotechnological applications. Frontiers in Plant Science, [online] 3. Available at: <https://www.frontiersin.org/articles/10.3389/fpls.2012.00222/full> [Accessed 2 July 2022].
6. Koivunen, M. and Krogsrud, R., 2022. Principles of Immunochemical Techniques Used in Clinical Laboratories. [ebook] p.8. Available at: <https://www.dbt.univr.it/documenti/Avviso/all/all226092.pdf> [Accessed 2 July 2022].
7. Nakai, Y. and Maruyama-Nakashita, A., 2020. Biosynthesis of Sulfur-Containing Small Biomolecules in Plants. International Journal of Molecular Sciences, [online] 21(10), p.3470. Available at: <https://www.ncbi.nlm.nih.gov/pmc/articles/PMC7278922/> [Accessed 2 July 2022].
8. O. Bruce, S. and A. Onyegbule, F., 2021. Biosynthesis of Natural Products. Biosynthesis [Working Title],
9. Priyadharshini, K., 2022. [ebook] Available at: <https://webstor.srmist.edu.in/web_assets/srm_mainsite/files/downloads/Immunochemical_Techniques.pdf> [Accessed 2 July 2022].
10. Priyadharshini, K., n.d. IMMUNOCHEMICAL TECHNIQUES. [ebook] Available at: <https://webstor.srmist.edu.in/web_assets/srm_mainsite/files/downloads/Immunochemical_Techniques.pdf> [Accessed 2 July 2022].

CHAPTER 3

GENETIC INFORMATION TRANSFER

CONTENTS

Genetic information, in general, provides continuity of life and thus it is really important to be clear about its process or mechanisms, in order to understand how it actually flows from one micro-organism to another. The chapter starts with the basic outline of genetic information transfer. After that, it talks about the processes of transformation, transduction and conjugation as a part of the genetic information transfer mechanisms. The complete process of transcription and translation has also been described. Towards the chapter's end, how genetic information exactly flows and the key notion of horizontal gene transfer have been explained in detail.

3.1. INTRODUCTION

Genetic information ensures life's continuance, so in most circumstances, this data is handed down from parents to children via DNA. Heritable information is packed into chromosomes which are transmitted to daughter cells through mitosis, a regulated system that assures each daughter cell obtains an identical and complete complement of chromosomes.

The generation of identical offspring allows organisms to develop, replace cells, and reproduce asexually. Sexual reproduction, on the other hand, includes the recombination of heritable information from both parents via the fusing of gametes during fertilization. Natural selection acts on a range of potential phenotypes in progeny produced by meiosis and fertilization.

Variability is ultimately caused by changes in genetic information. Adaptation to changing settings is impossible without variety. Some phenotypes are the result of single gene activity, and these features offered the experimental setup that Mendel used to establish a model of inheritance. However, scientists quickly discovered that standard Mendelian genetics did not explain the inheritance patterns of many features and that mutations in DNA may be useful tools for studying how genes are transferred or how genes operate.

A white-eyed male fruit fly (*Drosophila melanogaster*) mutation discovered in a culture of red-eyed fruit flies led to the discovery of the chromosomal basis of inheritance, the underlying explanation that evaded Mendel. Nowadays, biotechnology allows people to directly create heritable alterations in cells to produce new protein products.

Many methods exist for bacterial genetic exchange. The receptive bacteria absorb extracellular donor DNA during transformation. Donor DNA packed in a bacteriophage infects the recipient bacteria during transduction. Through mating, the donor bacteria transmit DNA to the receiver. The rearranging of donor and recipient genomes to generate new, hybrid genomes are known as recombination. Transposons are movable DNA segments that transfer from one genome to the next.

The biological importance of sexuality in microorganisms is that it increases the likelihood that uncommon, separate mutations may coexist in a single bacterium and also be exposed to natural selection. Microbe-microbe genetic interactions allow their genomes to develop considerably faster than mutation itself.

The fast appearance and spread of antibiotic resistance plasmids, flagellar phase variation in Salmonella, and antigenic variation of surface antigens in Neisseria and Borrelia are examples of medically significant occurrences involving genetic information exchanges or genomic rearrangements.

Reproductive activities in bacteria entail the transfer of genetic information from a source to a recipient, which results either in source allele replacement or the introduction of donor genetic material to the recipient genome. Transformation, transduction, and conjugation are sexual processes in which donor DNA is introduced into recipient bacteria in various ways.

Since donor DNA cannot survive in the recipient bacteria unless it is part of a replicon, recombination between the donor and recipient genomes is frequently necessary to create stable, hybrid offspring. When the donor and recipient bacteria are from the same or closely related species, recombination is more likely to happen.

To be identified, a recombinant's phenotype must be distinct from both parental phenotypes. Before the recombinant phenotype is manifested, growth or cell division may be necessary. Segregation lag refers to the delay in expression of a recombinant phenotype until a haploid recombinant genome has segregated, whereas phenotypic lag refers to the delay in the synthesis of products encoded by donor genes.

When the parent's bacteria have distinct alleles for numerous genes, testing for linkage (nonrandom reassortment of parental alleles in recombinant offspring) is possible. If an unselected gene is related to the selected donor gene, the donor allele is more likely to be present in recombinants than if it is not linked to the selected donor gene.

The development of genetic maps is made possible by quantitative linkage analysis. The genome of *E. coli* is circular, as determined by both genetic linkage and direct biochemical examination of chromosomal DNA, and the genetic map is colinear with the physical map of chromosomal

DNA. Extrachromosomal replicons like bacteriophages and plasmids are also studied via genetic and physical mapping.

Several bacteria have restriction-modification (RM) systems, which comprise modifying enzymes that methylate adenine or cytosine residues at specific sequences in their very own DNA and corresponding restriction endonucleases that cleave foreign DNA that does not transport the specific modification at the same target sequences.

Certain restriction enzymes would only break DNA that has been methylated at specified regions. Such restriction systems, which may have evolved to defend bacteria against invasion by phages or plasmids, are a key barrier to genetic exchanges between different bacterial strains or species. Recent research shows that plasmid-borne RM systems may be a means for the plasmid to secure its place in a host strain because cells that lack the plasmid (and the related protective methylase gene) are destroyed by the action of the much more persistent restriction enzyme, that targets freshly reproduced yet unaltered chromosomal DNA.

Bacterial horizontal gene transfer occurs via three mechanisms: transformation, transduction, and conjugation. Conjugation is the most prevalent process for horizontal gene transfer across bacteria, particularly from a donor bacterial species to distinct receiver species. Although bacteria can acquire additional genes through transformation and transduction, this is often a more infrequent transfer between bacterium of the same or closely linked species.

3.2. TRANSFORMATION

Pieces of DNA produced by donor bacterium are picked up directly from the extracellular environment by receiver bacterium during transformation. Recombination happens between single molecules of transformed DNA and receiving bacterial chromosomes. DNA molecules must be at least 500 nucleotides long to be active in transformation, and transforming activity is swiftly eliminated by treating DNA with deoxyribonuclease.

Transforming DNA molecules are very tiny pieces of the bacterial chromosome. Gene co-transformation is therefore improbable unless they are so closely related that they may be expressed on a single DNA fragment. Transformation was identified in *Streptococcus pneumoniae* and is also seen in Haemophilus, Neisseria, Bacillus, and Staphylococcus.

The capacity of bacteria to take up extracellular DNA and get converted, known as competence, changes with the organism's physiologic condition.

Many bacteria that are normally incapable of taking up DNA can be coaxed to do so using laboratory manipulations such as calcium shock or exposure to a high-voltage electrical pulse (electroporation).

DNA absorption in some bacteria (including Haemophilus and Neisseria) is dependent on the presence of particular oligonucleotide sequences in the transforming DNA, but not in others (including *Streptococcus pneumoniae*). Competent bacteria can also accept intact bacteriophage DNA or plasmid DNA, which can then reproduce as extrachromosomal genetic components in the recipient bacterium. Unless it becomes part of a replicon by recombination, a portion of chromosomal DNA from a donor bacterium normally cannot reproduce in the receiving bacterium. Traditionally, the characterization of the "transforming principle" of *S. pneumoniae* offered the first direct proof that DNA is the genetic material.

3.3. TRANSDUCTION

Bacteriophages operate as vectors in transduction, introducing DNA from donor bacteria into recipient bacteria via infection. A tiny percentage of the virions generated during lytic growth by some phages, known as generalized transducing phages, are aberrant and include a random piece of the bacterial genome instead of phage DNA.

Each transducing phage carries a unique collection of closely connected genes that represent a small portion of the bacterial genome. Generalized transduction is mediated by populations of such phages because each region of the bacterial genome has roughly the same likelihood of being transmitted from donor to recipient bacterium.

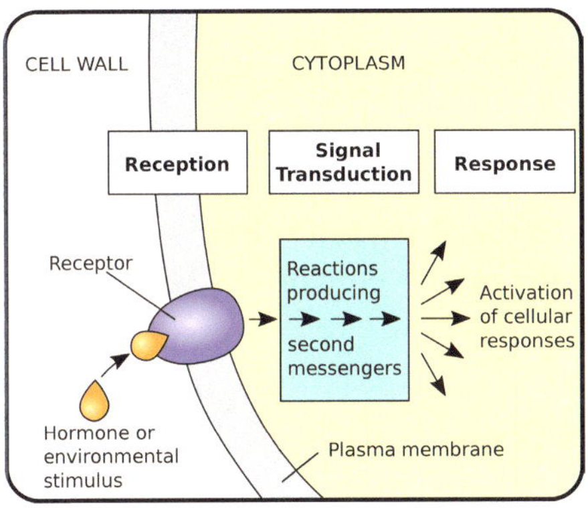

Figure 3.1. Transduction pathways.

Source: Image by Wikimedia

If a generalized transducing phage infects a recipient cell, the donor genes that have been transmitted are expressed. Abortive transduction is defined as the transitory expression of one or more donor genes without the generation of recombinant offspring, whereas full transduction is defined by the development of stable recombinants that inherit donor genes and maintain the capacity to express them.

The donor DNA fragment does not replicate in abortive transduction, and only one bacterium from the initial transductant has the donor DNA fragment. After each generation of bacterial development, the donor gene products get progressively diluted in every other progeny until the donor phenotype could no longer be reproduced.

Abortive transductants form minute colonies that are easily separated from stable transductant colonies on selective media that only bacteria with the donor phenotype can grow on. The incidence of abortive transduction is often one to two orders of magnitude larger than the frequency of generalized transduction, indicating that most cells infected by generalized transducing phages do not create recombinant offspring.

In various respects, specialized transduction varies from generalized transduction. It is exclusively mediated by temperate phages, and only a few donor genes may be transmitted to recipient bacteria. Only when lysogenic donor bacteria enter the lytic cycle and release phage offspring do specialize transducing phages arise.

Transducing phages are uncommon recombinants that lack a portion of the regular phage genome and have a portion of the bacterial chromosome close to the prophage attachment site. Many specialized transducing phages are dysfunctional and cannot complete the lytic cycle of phage development in infected cells without the presence of helper phages to perform missing phage functionalities.

The specialized transducing phage lysogenizes the recipient bacteria and expresses the donor genes, resulting in specialized transduction. Although there are many parallels between phage conversion and specialized transduction, the origin of the converting genes in temperate converting phages is unclear.

3.4. CONJUGATION

Direct contact between both the donor and recipient bacteria results in the formation of a cytoplasmic bridge and the transfer of part or all of the donor genome to the recipient. Specific conjugative plasmids known as fertility plasmids or sex plasmids determine donor ability.

The F plasmid (also known as the F factor) of *E. coli* serves as the model for fertility plasmids in Gram-negative bacteria. *E. coli* strains containing an extrachromosomal F plasmid are designated F+ and act as donors, whereas strains without the F plasmid are called F– and act as receivers. The conjugative capabilities of the F plasmid are defined by a cluster of at least 25 transfer (TRA) genes that control F pili expression, DNA synthesis and transfer during mating, interference with the capacity of F+ bacteria to serve as receivers, and other tasks. To commence mating, each F+ bacterium possesses 1 to 3 F pili that attach to a particular outer membrane protein (the ompA gene product) on the recipient bacteria.

A cytoplasmic bridge is created, and one strand of F plasmid DNA is transferred from donor to recipient, starting at a distinct origin and proceeding in the 5′ to 3′ direction. In the recipient bacteria, the transferred strand is transformed into circular double-stranded F plasmid DNA, and a new strand is generated in the donor to replace the transferred strand.

Because both of the exconjugant bacteria are F+, the F plasmid can propagate through infection across genetically compatible bacterial populations. In addition to their involvement in conjugation, the F pili act as receptors for donor-specific (male-specific) phages.

In *E. coli*, the F plasmid may persist as an extrachromosomal genetic element or be incorporated into the bacterial chromosome. Because the F plasmid and the bacterial chromosome are both circular DNA molecules, reciprocal recombination between them results in a bigger DNA circle composed of F plasmid DNA inserted linearly into the chromosome.

E. coli has numerous copies of several distinct genetic elements known as insertion sequences (for more information, see the section on transposons). Homologous recombination between insertion sequences in the chromosome and the F plasmid results in preferential integration of the F plasmid at chromosomal locations with insertion sequences. The chromosomal loci where insertion sequences are identified, though, differ amongst *E. coli* strains.

3.5. GENETIC INFORMATION IN MICROBES

DNA is the genetic substance found in bacteria and plasmids. Bacterial viruses (bacteriophages or phages) have DNA or RNA as genetic material. The two most important roles of genetic material are replication and expression. The genetic material must replicate correctly so that children inherit all of the particular genetic determinants (the genotype) of the parental organism. The observable traits (phenotype) of the organism are determined by the expression of certain genetic material under a given set of growth circumstances.

Bacteria have few clearly seen structural or developmental traits, but they have a wide range of metabolic capabilities and patterns of sensitivity to antimicrobial drugs or bacteriophages. These latter qualities are frequently chosen as the hereditary traits to be studied in bacterial genetics investigations.

3.6. TRANSCRIPTION / DNA TRANSCRIPTION

Transcription is the process of copying information from a strand of DNA into a new molecule of messenger RNA (mRNA). DNA maintains genetic material in cell nuclei as a reference, or template, in a secure and stable manner. Meanwhile, mRNA is analogous to a reference book copy since it has the same information as DNA but is not employed for long-term storage and may easily escape the nucleus. Although the mRNA includes the same information as the DNA segment, it is not an identical duplicate since its sequence is complementary to the DNA pattern.

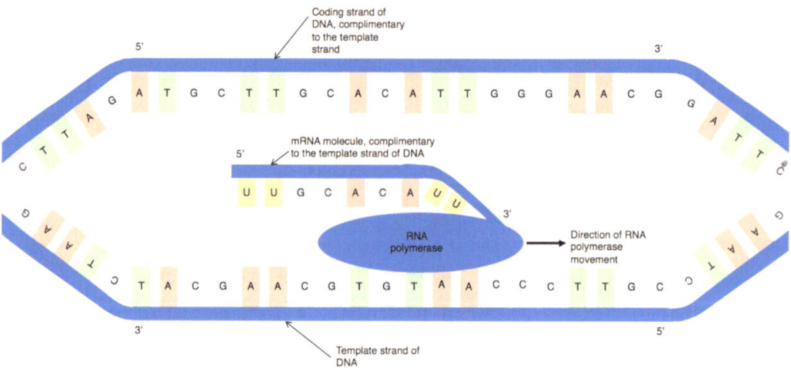

Figure 3.2. DNA transcription process.

Source: Image by Wikimedia

Transcription is carried out by an enzyme known as RNA polymerase and a group of proteins known as transcription factors. Transcription factors can bind to particular DNA regions known as enhancer and promoter sequences to attract RNA polymerase to a specific transcription site. The transcription factors and RNA polymerase work together to produce the transcription initiation complex.

This complex starts transcription, and RNA polymerase starts mRNA synthesis by matching complementary nucleotides to the original DNA strand. The mRNA molecule is stretched, and transcription is stopped after the strand is fully produced. During the translation process, the freshly generated mRNA copies of the gene serve as blueprints for protein synthesis.

Transcription is the initial stage of gene expression, in which information from a gene is used to build a functional output like a protein. The purpose of transcription is to create an RNA copy of the DNA sequence of a gene. The RNA copy, or transcript, of a protein-coding gene, contains the information required to construct a polypeptide (protein or protein subunit). Before even being translated into proteins, eukaryotic transcripts must go through various processing stages.

3.6.1. Stages of Transcription

A gene's transcription occurs in three stages: start, elongation, and termination. In this section, we will look at how these stages occur in bacteria. In the steps of transcription, you may learn more about the specifics of each stage (as well as how eukaryotic transcription differs).

- **Initiation**: RNA polymerase attaches to a stretch of DNA known as the promoter, which is present near the start of a gene. In bacteria, each gene (or collection of co-transcribed genes) has its own promoter. RNA polymerase, once attached, splits the DNA strands, giving the single-stranded template required for transcription.
- **Elongation**: The template strand of DNA acts as a guideline for RNA polymerase. The polymerase constructs an RNA molecule out of complementary nucleotides as it "reads" this template one base at a time, forming a chain that develops from 5' to 3'. The RNA transcript includes the same information as the non-template (coding) strand of DNA, but instead of thymine, it contains the nucleotide uracil (U) (T).
- **Termination**: Terminators are patterns that indicate the end of an RNA transcript. These enable the transcripts to be released from the RNA polymerase after it has been transcribed.

3.6.2. RNA Modification

Before it is translated into a protein, RNA must be changed once it is created from the DNA template. A cap is first put to the end of the RNA. This cap is a chemically modified nucleotide that aids in translation regulation. A poly-A tail is then appended to the end of the RNA. This tail is made up of numerous A nucleotides that are attached to the end of the RNA.

The tail's function is to protect the RNA molecule from exonuclease destruction. Although some eukaryotic chromosomal DNA contains information required to code for proteins, the vast majority of DNA does not contain such information. The coding DNA is referred to as exons, whereas the noncoding DNA is referred to as introns.

Unfortunately, the introns are situated within the exons, causing their sequence to be disrupted. Introns were classified as "junk DNA" for many years since their roles were unknown. Introns now have known roles like as controlling translation, serving as mobile genetic elements, and allowing for alternative splicing patterns. Intron acquisition appears to be a crucial aspect of eukaryotic evolution.

The sequences of both the introns and the exons are copied from the DNA during transcription. To generate functional mRNA, introns must be deleted and exons must be spliced together prior to translation. Spliceosomes are used to do this. Each spliceosome is made up of proteins known as small nuclear ribonucleoproteins (snRNPs) and small nuclear RNAs (snRNAs), which are responsible for exon splicing. SnRNAs are catalytic RNA molecules that are examples of ribozymes. By splicing the same exons in various sequences, several distinct RNAs can be created. Shows the RNA splicing process. After splicing, the mRNA molecule passes via the nuclear pores to the cytoplasm, where translation occurs.

3.6.3. The Genetic Code

After the mRNA is created, it must be translated into a protein. There is a "linguistic" barrier in this scenario. mRNA is encoded with a four-letter code (A, U, C, and G), whereas proteins are encoded with a twenty-letter code (there are 20 different amino acids used to make proteins). So, how can a four-letter language become a twenty-letter language?

The mRNA is read in three-nucleotide chunks called codons. Each codon has enough information to designate one amino acid. Mathematically, there are four nucleotides in mRNA, and if every three-letter combination is used, there are 64 potential codons, all of which are specified in the genetic code. Due to the fact that there are only 20 amino acids utilized to form proteins, there is overlap in the code, or redundancy, in which more than one codon can code for the same amino acid.

Understanding the order of codons on thc mRNA allows you to utilizc the genetic code to interpret the order of amino acids which will be used to form the protein during translation. Any modification to the DNA, which alters the mRNA codons, has the ability to modify the order of amino acids and consequently the structure and function of the intended protein.

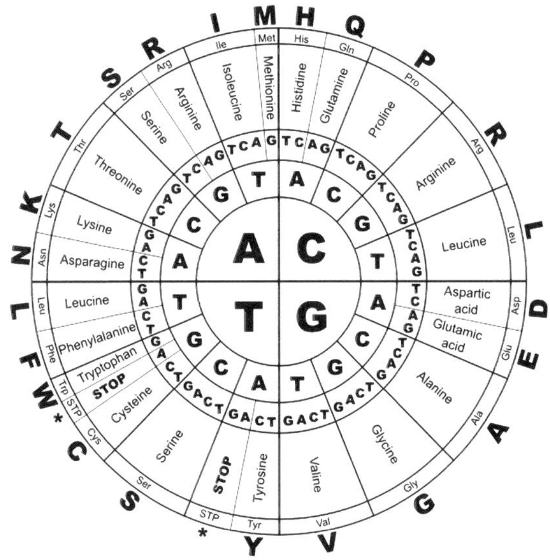

Figure 3.3. Genetic codes.

Source: Image by FreeSVG

3.6.4. Transfer RNA

The tRNA molecules transport the relevant amino acids to the ribosomes in accordance with the mRNA codons. The tRNA is a fragment of RNA that has been folded into a particular structure. The anticodon on one end of the tRNA is complementary to the codon on the mRNA. For example, if the mRNA codon is CAU, the anticodon on the tRNA is GUA.

A particular amino acid is connected to the opposite end of the tRNA. The aminoacyl-tRNA synthetases are in charge of connecting amino acids to the appropriate tRNA. According to the genetic code, there are 64 codons, three of which serve as stop codons, leaving 61 functioning codons. This implies that 61 different tRNAs containing anticodons are required to match these mRNA codons.

This, however, is not the case. Wobble pairing occurs in tRNAs, when the third base of the anticodon can bind to the third base of the codon in a noncomplementary manner. A tRNA with the anticodon UAG, for example, would be anticipated to bind to a codon with the entire complementary sequence AUC. However, the identical UAG anticodon on the tRNA might bind to a codon with the pattern AUU, where the third base is not a normal complementarity match.

3.7. TRANSLATION

Translation takes place in the cytoplasm. The codons on the mRNA are read, and the amino acids required to make the protein are assembled. Various enzymes, ribosomes, and tRNA are required to aid in this process.

Ribosomes

Eukaryotic ribosomes are made up of two subunits of rRNA, one big and one tiny, as well as other proteins. When the ribosome assembles on the mRNA, it has two RNA binding sites: the peptidyl (P) site and the aminoacyl (A) site.

3.7.1. Transfer RNA

The tRNA molecules transport the relevant amino acids to the ribosomes in accordance with the mRNA codons. The tRNA is a fragment of RNA that has been folded into a particular structure. The anticodon on one end of the tRNA is complementary to the codon on the mRNA. For example, if the mRNA codon is CAU, the anticodon on the tRNA is GUA. A particular amino acid is connected to the opposite end of the tRNA. The aminoacyl-tRNA synthetases are in charge of connecting amino acids to their respective tRNAs...

According to the genetic code, there are 64 codons, three of which serve as stop codons, leaving 61 functioning codons. This implies that 61 different tRNAs containing anticodons are required to match these mRNA codons. This, however, is not the case. Wobble pairing occurs in tRNAs, when the third base of the anticodon can bind to the third base of the codon in a noncomplementary manner.

A tRNA with the anticodon UAG, for illustration, would be anticipated to bind to a codon with the entire complementary sequence AUC. However, the identical UAG anticodon on the tRNA might bind to a codon with the pattern AUU, in which the third base is not a normal compatibility match.

3.8. THE STEPS OF TRANSLATION

The translation is a three-step procedure. Firstly, the ribosome must assemble on the mRNA. Secondly, the amino acids specified by the codons must therefore be transported to the ribosome and linked together in the direction of the N terminus to the C terminus. Lastly, the resulting protein must be released from the ribosome.

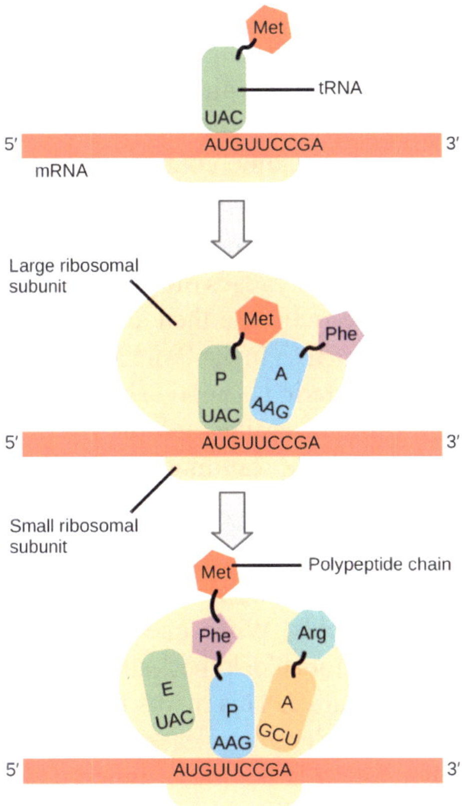

Figure 3.4. Schematic representation of translation process.

Source: Image by Wikimedia

3.8.1. Initiation

The translation process begins when the ribosome assembles on the mRNA and requires a number of starting factors. The start codon (AUG) on mRNA indicates the site of ribosomal assembly. The tiny ribosomal subunit then attaches to the mRNA. The initial tRNA binds to the P site of the ribosome. This tRNA must have the correct anticodon (UAC) to hydrogen bond with the start codon (AUG). The amino acid designated by the start codon, as shown in the genetic code, is methionine. Methionine will thus be the initial amino acid in every protein. The big subunit of the ribosome may now assemble on the mRNA.

3.8.2. Elongation

The P site of the ribosome is now occupied, while the A site is not. A tRNA with the right anticodon to interact with the mRNA's next codon will enter the ribosome and hydrogen bond to the codon. At this stage, a crucial enzyme will be employed to create a peptide connection between the two amino acids at the P and A positions. Peptidyl transferase is the enzyme in question.

The two amino acids are now bound to the tRNA at the A site. The tRNA in the P site migrates to the E site, breaks off (leaving its amino acid behind), and exits the ribosome. The ribosome then advances to the codon on the right, depositing the leftover tRNA in the P site and creating an empty A site. This cycle of a fresh tRNA entering, a peptide bond forming between amino acids, the tRNA at the P site departing, and the ribosome moving over by one codon will repeat indefinitely, resulting in a developing peptide.

3.8.3. Termination

Three mRNA codons (UAA, UAG, and UGA) function as stop codons and do not encode amino acids. When one of these codons reaches the ribosome's A site, release factors block the A site and the protein is released. Ribosomal subunits will separate. This marks the end of the translation. In certain circumstances, modifying the released protein is required before it may be used. Phosphorylation is the most common type of modification; however additional types of modification are conceivable. This procedure is common in eukaryotic cells' endoplasmic reticulum or Golgi complex.

3.9. GENETIC INFORMATION TRANSFER PROMOTES COOPERATION IN BACTERIA

Microorganisms release a plethora of components, including signaling, resource scavenging, pathogenicity, and anti-competitor compounds. These so-called "public good" molecules are expensive to make, but they are accessible and potentially useful not just to the organisms secreting them, but also to their neighbors.

Producing molecular public goods improves the fitness of nearby individuals and can thus be termed cooperative activity. The growth of individuals who gain from collaboration without paying the expenses, known as "nonproducers" or "cheaters," poses a broad threat to the continuation of cooperation. According to social evolution theory, cooperation can be sustained if its advantages are given preferentially to animals harboring cooperative genes.

Several methods have been hypothesized to bias the partner association in cooperative interactions, including the most current application of sociobiology ideas to microbes has enabled them to be tested experimentally. In particular, restricted dispersal of clone mates or homophilic receptor binding might result in a positive assortment among individuals bearing cooperative genes, which can be evaluated using relatedness statistics.

The frequent and unusual forms of sex in bacteria make their genomes extraordinarily plastic: genes often move within and across bacterial lineages, mostly through interaction with mobile genetic components such as plasmids or phages. Surprisingly, genes encoding cooperative features like as extracellular antibiotic degradation or cholera toxicity are frequently found on mobile elements, implying a relationship between social behaviors and horizontal transmission.

Two strategies can explain cooperative gene localization on mobile elements. First, despite the possible fitness cost to host cells, sufficiently large levels of horizontal transfer may promote mobile genetic elements merely as molecular parasites. Mobility may thus allow accessory genes to remain in the environment in the absence of ongoing positive selection.

Horizontal transmission of a cooperative producer allele might change a receiver from a nonproducer to a producer and therefore induce cooperation by infection in the case of public benefit encoding genes. Cooperation maintenance via infection enforcement, on the other hand, is projected to be unstable, as a noncooperative gene would still displace a cooperative allele when both move horizontally. Second, horizontal transmission has the potential to alter population-relatedness patterns (gene assortment).

Theoretical work shows that horizontal gene transfer will boost relatedness at mobile loci because the local dissemination of mobile alleles increases the likelihood that surrounding individuals will have the same allele, promoting investment in cooperative qualities at these loci. However, there is little scientific proof that transfer may appreciably alter the selection forces operating on cooperative behavior via either of these methods. For the first time, to the available knowledge, the consequences of horizontal transfer on collaboration experimentally have been studied.

In *Escherichia coli*, researchers created a synthetic system with separate regulations of cooperation and conjugation. The secretion of a public good, the *Pseudomonas aeruginosa* quorum-sensing molecule C4-HSL, is the cooperative feature. The chemical is expensive to make, but it promotes quicker growth in both producer and nonproducer cells when exposed to the antibiotic chloramphenicol (Cm) via triggering the expression of a chromosomal resistance gene.

Using the F plasmid's conjugation machinery, a helper F plasmid was created that allows the transfer of mobilizable (T+) plasmids with (producer, P+) or without (nonproducer, P) C4-HSL public good gene, however, does not, however, transmit non - mobilizable (T) plasmids. Initial plasmid-free cells (recipients, R), P+ and P plasmids are labeled with distinct fluorescent proteins to allow flow cytometry monitoring of strain and plasmid dynamics. This synthetic technique enables us to investigate the effect of horizontal transfer in a physiologically relevant situation free of plasmid–host coadaptation.

3.10. CELLULAR ORGANIZATION OF THE TRANSFER OF GENETIC INFORMATION

In eukaryotic cells, every step involves the transfer of genetic information is spatially controlled, as transcription, translation, and mRNA degradation all take place in discrete functional compartments (e.g., nucleus, cytoplasm and P-bodies). Because of the noticeable absence of membrane-enclosed organelles in bacteria, these activities appear to take place in the same compartment – the cytoplasm.

Moreover, it is becoming clear that mRNA-related activities are also spatially ordered inside bacterial cells, and that this organization influences cellular function. The goals of this section are to outline what is known about this structure and to look at the mechanisms and factors that shape the cell interior. The area is at an interesting stage when new technologies are allowing long-standing questions to be tested.

3.10.1. Two Types of Cellular Organization

Piekarski characterized a central DNA-rich zone, which he termed as nucleoid for "nucleus-like" in 1937, as the first clue of spatial order in the bacterial cell. Ribosomes preferentially collect outside this nucleoid, according to microscopy investigations. Sucrose gradient centrifugation investigations demonstrated that 50–85% of the ribosomal material in *E. coli* extracts elutes with the polyribosome fraction, implying that the majority of ribosomes are coupled to mRNA and involved in translation.

The majority of RNA polymerase (RNAP) molecules attach to the nucleoid, with approximately 7–14% remaining dispersed in the cytoplasm. Thus, separate zones of concentration of DNA (RNAP) and ribosomes suggest that the majority of transcription and translation is spatially segregated. Ribosome and DNA partitioning has been reported in a number of bacterial species and has become a classic depiction of bacterial cellular structure. Bacteria such as *Caulobacter crescentus*, *Agrobacterium tumefaciens*, *Sinorhizobium meliloti*, and certain mollicutes, on the other hand, represent an undervalued sort of organization.

The chromosomal DNA distributes across the cytoplasmic space in these species, and the ribosomes are uniformly distributed throughout the cytoplasm. This difference in cellular architecture illustrates that ribosomes and chromosomal DNA may coexist. It should be highlighted that, even in this instance, there is evidence that genetic information transmission is still subject to geographical restrictions.

3.10.2. Entropy and the Nucleoid Organizers

Which factors affect the geographic distribution of chromosomes and ribosomes? The bacterial cytoplasm has an extremely high mass density, with macromolecule concentrations exceeding 400 mg/ml in *E. coli*. As a result, the supercoiled chromosome is exposed to crowding pressures that cause the DNA to compress.

Mondal et al. used a Monte Carlo simulation based on experimentally established assumptions to show that the separation of ribosomes and DNA in *E. coli* may be explained by maximizing the chromosome's conformational entropy and the translational entropy of ribosomes. While this attractive model outlines key entropic forces to consider, additional variables are most likely at work. For example, this physical model does not explain how certain bacteria have an alternative cellular arrangement in which ribosomes and DNA are present.

Furthermore, it has been demonstrated that intra-chromosomal interactions influence chromosome architecture. As recently proven by atomic force microscopy, super-resolution fluorescence microscopy, and chromosome conformation capture analysis, nucleoid-associated proteins (NAPs) bind, bend, coat, and bridge DNA segments, therefore engaging in the chromosomal organization.

Thus, the spatial distribution of chromosomes and ribosomes is most likely determined by a balance of physical restrictions and biological interactions that modify the physical characteristics of DNA. A shift in equilibrium may result in a shift in cellular structure. The different shapes and orientations of the chromosomes of *E. coli* and *C. crescentus* point to this idea.

3.10.3. Spatially Separated, yet Functionally Coupled

In *E. coli*, the segregation of ribosomes and DNA (and RNAP) suggests that transcription and translation are spatially segregated. In the same bacteria, however, a functional connection between translation and transcription (i.e., translation begins on nascent mRNAs as they are transcribed) has been extensively demonstrated.

The concept of coupling was suggested in 1964 and received experimental confirmation in the early days of molecular biology. A transcribing RNAP's physical contact with a trailing ribosome has been lately detailed in atomic detail. How can the seeming contradictions be reconciled? We anticipate at least three eventualities that are not mutually exclusive.

3.10.4. RNA Degradosome Localization

Whilst transcription and translation take the majority of the focus, mRNA degradation is a critical mechanism in the flow of genetic information because it influences the levels (and presumably the location) of mRNAs.

Most gene expression models assume that mRNA degradation occurs equally across the cytoplasm.

Recent research on several components of the RNA degradosome, however, suggests otherwise. In *E. coli*, RNase E, a key component of the mRNA degradation process, has been found to bind with the membrane. RNase Y (YmdA), the functional counterpart of RNase E in *B. subtilis*, was likewise discovered to be membrane-associated. Furthermore, other RNA degradosome components (for example, RNA helicase B, polynucleotide phosphorylase, and enolase) have non-homogeneous distributions inside cells. Surprisingly, a fluorescent RNase E fusion shows no membrane accumulation in *C. crescentus*, instead of displaying a patchy pattern across the cell.

More notably, in filamentous cells containing DNA-free areas, this fusion segregates with the chromosome. This shows that *C. crescentus* RNase E interacts with chromosomal DNA (directly or indirectly). Alternatively, its distribution may mirror the distribution of (slowly diffusing) mRNAs. In any case, the variation in RNase E localization patterns between *C. crescentus* and *E. coli* highlights the issue of whether it is a source or a result of the two forms of bacterial cellular structure.

3.11. THE FLOW OF GENETIC INFORMATION

The basic function of DNA in bacteria, archaea, and eukaryotes is to store heritable information that encodes the instruction set necessary to create the organism in question. While we've gotten much better at quickly reading the chemical composition (the sequence of nucleotides in a genome and some of the chemical modifications made to it), researchers still don't understand how and where to dependably decode all of the details within including all the mechanisms with which it is read and eventually expressed.

Yet, there are several key concepts and mechanisms connected with the reading and expression of the genetic code whose basic stages are recognized and that should be part of any biologist's conceptual arsenal. Transcription and translation, which are the coping of sections of the genetic code inscribed in DNA into molecules of the similar polymer RNA and the reading and encoding of the RNA code into proteins, are two of these processes.

We next detail how Nature uses the process to generate a range of functional RNA molecules (with varied structural, catalytic, or regulatory activities), including messenger RNA (mRNA) molecules that transport the information needed to make proteins. Similarly, we concentrate on the difficulties and questions that arise throughout the translation process.

In biological systems, the basic flow of genetic information is sometimes portrayed in a framework known as "the core dogma". According to this model, information encoded in DNA is translated into RNA and then into proteins via translation. Reverse transcription (the production of DNA from an RNA template) and replication are further processes for information propagation in various ways.

This system, however, says nothing about how information is encoded or the methods by which regulatory signals pass across the model's numerous levels of molecule kinds. As a result, while the system below is a nearly essential element of every biologist's lexicon, probably inherited from old tradition, readers should also be informed. (That information flow processes are more complicated, and that "the central dogma" just covers certain core routes).

3.12. HORIZONTAL GENE TRANSFER

Horizontal gene transfer (HGT) is the steady transfer of genetic material from one organism to another that does not need reproduction. The importance of horizontal gene transfer was initially recognized when evidence of 'infectious heredity' of various antibiotic resistance to organisms was discovered. HGT's supposed importance has shifted multiple times, although there is currently a widespread consensus that HGT is a substantial, though not dominant, driver in bacterial evolution.

Deviant composition, aberrant phylogenetic distribution, the high closeness of genes from distantly related species, and incongruent phylogenetic trees revealed massive gene transfers in fully sequenced genomes. There is also a growing body of evidence that HGT by mobile genetic elements (MGEs) is a continuous process that plays a key role in prokaryotic ecological adaptability.

The spread of antibiotic resistance genes through HGT is well known, allowing bacterial populations to quickly adjust to a high selective pressure by agronomically and medically used antibiotics. MGEs shape bacterial genomes, increase intra-species diversity, and disperse genes among distantly related bacterial genera.

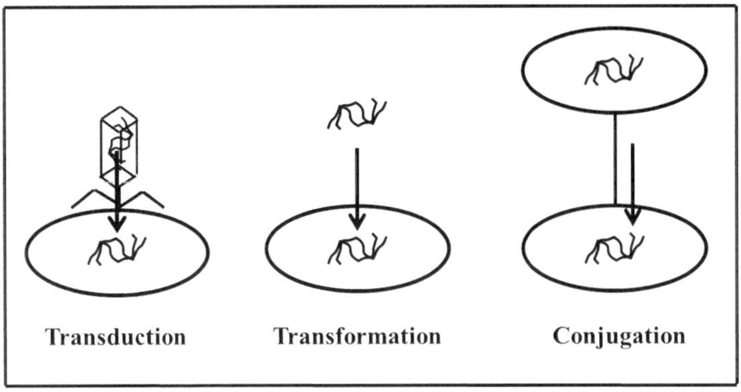

Figure 3.5. Gene transfer in bacteria.

Source: Image by Wikimedia

Implications of horizontal gene transfer to risk assessment of transgenic microorganisms

The presence of horizontal gene transfer among bacteria has significant implications for assessing the danger of transgenic microorganisms. Because horizontal gene transfer between bacteria is reported to exist in the ecosystem, it is necessary to assess the possibility for future gene transfer of inserted genetic sequences from a transgenic bacterium to indigenous microorganisms when that bacterium is discharged into the environment.

Transgenic bacteria, like their unmodified parent organism, may collect mobile genetic elements in the environment, which may increase their tolerance to environmental challenges and therefore their fitness. There are two key principles in assessing the danger of gene transfer from transgenic microorganisms. The first is a risk element that explores both possibility of transmission as well as the prospective extent (i.e., range of recipient organisms) of gene transfer from the transgenic micro-organism to other micro-organisms in the environment as a result of the transgenic micro-unique organism's design or construction, taking into account the parental micro-organism, the introduced genes, the method(s) of insertion, as well as the environmental habitat into which the GEM is introduced.

The examination of potentially unfavorable consequences that may result from gene transfer is another crucial component in the risk assessment of transgenic microorganisms. The assessment of the potential effects of gene transfer is the most essential factor in an examination of the hazards involved with gene transfer, and hence will be the focus of this section first.

3.12.1. Potential Adverse Effects of Horizontal Gene Transfer from a Transgenic Bacterium to Indigenous Micro-Organisms

The danger of horizontal gene transfer of transgenes from a transgenic bacterium is contingent on the possibility of any detrimental consequences caused by the transgenes' uptake in other microorganisms. If any phenotype associated with the inserted gene(s) is potentially undesirable in another host background, such as toxicity, pathogenicity, increased virulence, antibiotic resistance, competitive advantage, use of novel substrates, or vastly expanded host range, then the ability for gene transfer should be closely examined.

On the other hand, if the inserted gene(s) of interest do not have any deleterious effects or perform any unique functions, assessing the risk exposure components becomes less significant. Similarly, if there is a big existing gene pool for genes in the environment, Similarly, if there is already a significant existing gene pool in the atmosphere for genes trying to impart a specific inserted trait (e.g., degradative pathways such as SAL, TOL, and NAH), there may be little cause for concern, even if gene transfer from the transgenic bacterium to other microorganisms in the environment occurs easily.

Likelihood and extent of gene transfer

The manner by which a transgenic bacteria is constructed may impact the chance of gene transfer from that bacterium to other microorganisms in the environment. The possibility of transmission by any of the bacterial horizontal gene transfer processes (i.e., conjugation, transformation, and transduction) bears attention in this analysis.

Genes introduced using plasmid vectors

The use of plasmids as carriers that remain solid extrachromosomal elements in the transgenic bacterium might enable horizontal gene transfer from that bacterium to certain other bacteria, including subsequent expression of gained genes. As with plasmids in naturally found bacteria, the possibility and scope of the transfer of a plasmid used as a vector to produce a transgenic bacterium are dependent on a number of plasmid-specific factors, like host range, the existence of resident plasmids in a potential recipient, and character traits of potential environmental recipients, as previously discussed in this document.

Evaluations of the possibility and amount of plasmid vector transfer from the transgenic microorganism to indigenous microorganisms should take into account. if the plasmid in the final transgenic microorganism construct contains an undamaged mobilisation (nick) site and transfer (TRA) genes A conjugative plasmid may become nonself-transmissible if these genes are disrupted. However, the prospect of missing TRA functions being provided in trans by another self-transmissible plasmid must be considered.

The use of plasmid vectors for gene insertion into transgenic microorganisms may result in the transmission of those genes via various horizontal gene transfer processes like as transformation and transduction. However, the frequency of plasmid transfer via these methods is likely to be significantly lower than that predicted by the conjugation of self-transmissible plasmids.

The size of the plasmid vector may affect its capacity to be transmitted through transformation, as bigger plasmids are less likely to remain intact for lengthy periods of time in an environment where DNA is prone to destruction by nucleases. Similarly, the size of nucleic acid sequences that transducing phages may carry is physically limited. Nonetheless, these horizontal gene transfer methods must be addressed in the biosafety assessment of modified organisms.

Genes inserted into the bacterial chromosome

Gene transfer from transgenic microorganisms created by steady integration of the inserted genes of interest into the recipient bacterial chromosome may not be quite as problematic as gene transfer from conjugative plasmid vectors. Chromosomal insertion has been performed effectively via a number of approaches, including the use of suicide plasmids or conditional replicons, as well as transposon activity.

Other processes, like transduction or transformation, or artificial transformation techniques, such as electroporation or the use of a gene gun, can also be used to insert genes into a recipient's chromosome. Regardless of the technique, the imported DNA gets incorporated into the chromosome at low rates by homologous recombination or perhaps even illegitimate/non-homologous recombination.

While chromosomal integration may lower the frequency or likelihood of horizontal gene transmission from the transgenic bacterium to other bacteria, gene transfer remains feasible. Some places, such as genomic islands, may be naturally more mobile. Transposition can occur, leading the inserted gene to become unstable from one location on the chromosome to another, or even to distinct replicons within a cell.

The presence of genomic islands (including pathogenicity islands) with independent movement via genes generating integrases and transposons in microbial populations underlines the evolutionary significance of these events. These islands are recognized vectors for gene transfer from chromosomal regions. They are commonly viewed as mobilizing components of the flexible gene pool, as opposed to "core genes," which frequently code for vital tasks and appear to be less susceptible to mobilisation. Transducing phages can also transmit chromosomal genes.

Transformation of chromosomal DNA fragments from dead cells is still possible. Although the incidence of chromosomal gene transfer is assumed to be lower than that of genes carried on self-transmissible plasmids, risk evaluation must take into account the mobile elements located within the chromosome, which may cause instability and therefore transfer.

3.12.2. Other Factors Influencing Horizontal Gene Transfer

Aside from the distinct building methods and ultimate position of introduced genes inside the genome of a transgenic bacterium, there seem to be a number of factors that determine the possibility and amount of horizontal gene transfer of inserted genes to indigenous microorganisms.

These include the gene itself, along with its phenotypic trait, especially if it confers a selective advantage to a beneficiary, the inherent capacity of the parental bacterium to transfer genes in general, the capacity of the transgenic bacterium to stay alive in the environment, the existence of appropriate recipient bacteria in the environment, as well as different environmental variables that may impact microbial activity and transfer such as moisture content, nutrient status, clay content, and so on.

Until properly modified for HGT, the most typical experience with utilizing transgenic microorganisms in laboratory and field testing is a failure to retain their existence or transmit the inserted genes. As previously stated, the characteristic conferred by a gene(s) is vital in determining whether gene transfer is possible. Irrespective of genomic location, genes encoding features imparting a selection advantage are likely to spread and establish in bacterial populations.

The presence of genomic islands giving qualities such as pathogenicity, symbiosis, fitness, or resistance in microbial populations indicates that such characteristics that are favorable or helpful to the host are widely conveyed. Novel genes that give no competitive benefit to a new host, on the other hand, may be lost or not expressed.

Certain transgenic microorganisms have been demonstrated to exhibit worse fitness when compared to their unaltered parental strains, which in some circumstances could be linked to a metabolic depletion of the cell. In those other circumstances, certain transgenic microorganisms have shown lower fitness because of the cytotoxic effects of added genes. Because gene transfer is influenced by the physiological condition of the cell, the gene itself may be helpful in determining gene transfer.

The capacity of the parental bacteria used to create a transgenic microorganism to exchange genes may impact the possibility of horizontal gene transfer. Natural gene exchangers are bacterial species found in the environment. Pseudomonads, for example, are infamous for the transmission of degradative genes (usually found on plasmids) across members of different genera.

Closely related taxa, such as Pseudomonas and Burkholderia, may easily interchange genes, most likely due to the presence of the necessary host machinery for both gene transfer and expression. Broad host range plasmids (BHR) such as IncP-1 plasmids, as well as newly found BHR groups like pIPO2), move easily between distantly related bacteria.

It is critical to acknowledge the adaptability of bacterial genomes. Precise taxonomic identification of the receiver microbe utilized in the development of a transgenic microorganism might aid in determining the capacity for gene transfer from the transgenic microorganism to other microorganisms in the environment. The significance of using appropriate phenotypic and genotypic tests and procedures for accurate bacterial identification has already been discussed.

The transgenic bacteria must live, at least temporarily, for numerous types of horizontal gene transfer to occur. If the transgenic microorganism is unable to survive in the environment, the possibility for transmission is considerably reduced, at least for conjugation and transduction pathways. There is still the possibility of the transformation of DNA released from dead cells, however, without bacterial replication, the number of cells would be restricted.

As previously stated, the longevity of naked DNA in the environment is reliant on a variety of circumstances, including adsorption to bacterial membranes, clay minerals, and other surfaces that give protection against nuclease breakdown. If the transgenic bacteria survive in the environment, the availability and concentration of suitable receivers influence the degree of possible gene transfer. As previously described in this paper, there are "hot spots" in soil and water habitats, as well as on plant surfaces, where microbial populations are higher, increasing the possibility for horizontal gene transfer, particularly by conjugation.

Quorum sensing systems, such as those seen in Agrobacterium, might also drive conjugation at greater cell densities. Similarly, horizontal gene transfer via transduction and transformation is more frequent in locations with higher microorganism densities. The potential for horizontal gene transfer from a transgenic microbe to other microorganisms in the environment is influenced by environmental conditions.

Abiotic parameters such as moisture content, temperature, nutritional status, and soil pH can all have an impact on the physiological condition of both the donor and recipient microorganisms. Furthermore, environmental factors may influence the establishment of transformation competence, and abiotic variables such as clay type and content or organic matter content may shield free nucleic acid sequence from degradation.

In situ accumulation of transportable genetic components that may impart selected features by the transgenic microorganism was demonstrated in two consecutive years for the transgenic *Pseudomonas fluorescens* SBW25EeZY6KX under field circumstances. Transconjugants with Hgr plasmids were recovered from the rhizosphere and phyllosphere at specific stages of plant growth.

A transgenic microorganism's properties may alter as a result of the acquisition of additional genetic information. The type and frequency of post-release genetic information acquisition will be determined by the environmental milieu into which the transgenic microorganism is discharged, as well as the host organism.

3.13. CONCLUSION

Changes in genetic information are the ultimate source of variability. Transformation, transduction, and conjugation are sexual processes that use different mechanisms to introduce donor DNA into recipient bacteria. Delay in the expression of a recombinant phenotype until a haploid recombinant genome has segregated is called segregation lag, and delay until synthesis of products encoded by donor genes has occurred is called phenotypic lag.

Genetic information provides for continuity of life, and, in most cases, this information is passed from parent to offspring via DNA. Recombination is most likely to occur when the donor and recipient bacteria are from the same or closely related species. Representative phenomena of medical importance that involve exchanges of genetic information or genomic rearrangements include the rapid emergence and dissemination of antibiotic resistance plasmids, flagellar phase variation in Salmonella, and antigenic variation of surface antigens in Neisseria and Borrelia. In transduction, donor DNA packaged in a bacteriophage infects the recipient bacterium. In transformation, the recipient bacterium takes up extracellular donor DNA.

The most common mechanism for horizontal gene transmission among bacteria, especially from a donor bacterial species to different recipient species, is conjugation. Recombination is the rearrangement of donor and recipient genomes to form new, hybrid genomes. Sexual reproduction, however, involves the recombination of heritable information from both parents through the fusion of gametes during fertilization.

There are three mechanisms of horizontal gene transfer in bacteria: transformation, transduction, and conjugation. Because donor DNA cannot persist in the recipient bacterium unless it is part of a replicon, recombination between donor and recipient genomes is often required to produce stable, hybrid progeny.

REFERENCES

1. 2022. Genetics and Information Transfer. [ebook] Available at: <https://apcentral.collegeboard.org/pdf/bio-bigidea3-geneticsandinformationtransfer.pdf?course=ap-biology> [Accessed 2 July 2022].
2. Biology LibreTexts. 2022. 17.1: The Flow of Genetic Information. [online] Available at: <https://bio.libretexts.org/Courses/University_of_California_Davis/BIS_2A%3A_Introductory_Biology_(Easlon)/Readings/17.1%3A_The_Flow_of_Genetic_Information> [Accessed 2 July 2022].
3. Campos, M. and Jacobs-Wagner, C., 2013. Cellular organization of the transfer of genetic information. Current Opinion in Microbiology, [online] 16(2), pp.171-176. Available at: <https://www.ncbi.nlm.nih.gov/pmc/articles/PMC3646911/>.
4. Dimitriu, T., Lotton, C., Bénard-Capelle, J., Misevic, D., Brown, S., Lindner, A. and Taddei, F., 2014. Genetic information transfer promotes cooperation in bacteria. Proceedings of the National Academy of Sciences, [online] 111(30), pp.11103-11108. Available at: <https://www.pnas.org/doi/10.1073/pnas.1406840111> [Accessed 2 July 2022].
5. Harmonisation of Regulatory Oversight in Biotechnology, 2010. Safety Assessment of Transgenic Organisms, Volume 4.
6. Holmes, R. and Jobling, M., 2022. Genetics. [online] Ncbi.nlm.nih.gov. Available at: <https://www.ncbi.nlm.nih.gov/books/NBK7908/> [Accessed 2 July 2022].
7. Nature.com. 2022. transcription / DNA transcription | Learn Science at Scitable. [online] Available at: <https://www.nature.com/scitable/definition/transcription-dna-transcription-87/#:~:text=Transcription%20is%20the%20process%20by,of%20messenger%20RNA%20(mRNA).> [Accessed 2 July 2022].

CHAPTER 4

CHROMATOGRAPHY AND BIOCHEMISTRY

CONTENTS

Chromatography is a technique that has numerous applications in biological as well as chemical fields. It is extensively used in biochemical research for the separation and identification of chemical compounds of biological origin. Hence, it is worthy to study the process in detail. The chapter starts by defining basic principles of chromatography including liquid and gas chromatography. Later, some of the performance parameters used in this process are explained. Equipment used in chromatography and different modes of it are described. Towards the end, the process of chromatography of a sample protein has been described.

4.1. INTRODUCTION

Hundreds of thousands of chemical species can be found within living cells. Large molecules like proteins, nucleic acids, and lipids, as well as lesser molecular weight compounds that serve as building blocks for biopolymers or components of complex metabolic pathways, fall into this category. Some of these compounds are found in minute amounts (for example, intermediates in enzyme processes), while others are abundant (e.g., structural proteins).

Furthermore, some components are only present during specific stages of the cell cycle, whilst others are present at nearly constant levels. Individual chemical components of cells can thus be studied to get insight into a variety of fundamental cellular processes and to better understand cell composition and function dynamics.

Separating individual chemical species using analytical or preparative chromatography is one method of studying them. Tswett (1903) first utilized this approach to separate plant pigments (chromatography originates from the Greek chroma, which means color), but we now know that it may be applied to any chemical species, colored or not.

Because biomolecules come in such a wide range of sizes, shapes, and hydrophobicity, no single chromatography approach will be adequate for all separations. The basic concepts of chromatography will be discussed in this chapter to explain why various molecules can be separated. Following that, some examples of the most common chromatographic procedures used in biochemistry are given to show how biomolecules are separated in biochemistry.

4.2. PRINCIPLES OF CHROMATOGRAPHY

4.2.1. The Partition Coefficient

A molecule may partition between the phases in any two-phase system (e.g., liquid–liquid, liquid–solid) (Figure 4.1). The precise ratio of concentration reached is ultimately controlled by the molecule's inherent thermodynamic qualities (which are a result of its chemical structure) and the phases' inherent thermodynamic properties. The relative solubility of the molecule in each liquid will be highly essential in determining partitioning in a liquid–liquid system. Different sample molecules may adsorb to varying degrees on the solid phase in a liquid–solid system. In a column system, both partition and adsorption phenomena can occur, which is referred to as column chromatography. In column chromatography, one phase (the stationary phase) is kept stationary while the other (the mobile phase) can freely flow over it. It can be expressed as the concentration ratio in such a system as the partition coefficient, K.4:

$$K = \frac{Cs}{Cm} \qquad \text{Eq (4.1)}$$

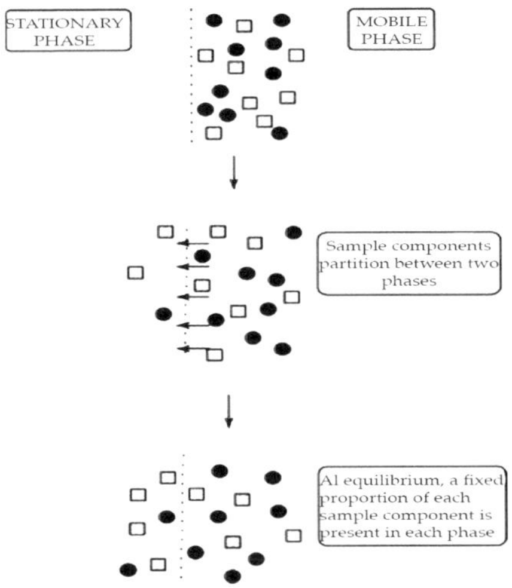

Figure 4.1. Partitioning of biomolecules in a two-phase system. Circles and squares, respectively, symbolize two components. An aqueous buffer and a solid stationary phase could be the two phases. In this experimental setting, the partition coefficients of the two samples are drastically different.

Source: Image by biologicaproceduresonline

As a result, each molecule will interact with the stationary phase in somewhat different ways and will migrate across the column at varying speeds due to different partitioning between phases. The chromatographer may alter the experimental parameters (e.g., temperature, solvent polarity) to maximize separation because K is directly affected by them.

To accomplish efficient separation in column chromatography, we use what are generally minor variances in the partitioning behavior of sample molecules.

4.2.2. Phase Systems Used in Biochemistry

The stationary phase in biochemistry chromatographic systems is made up of solid particles or solid particles covered with liquid. Chemical groups are frequently covalently connected to the particles in the former situation, which is referred to as bonded phase liquid chromatography. A liquid phase may coat the particle and be linked by noncovalent, physical attraction in the latter scenario. Liquid–liquid phase chromatography is the name for this type of system. Silica covered with a nonpolar hydrocarbon is a nice example of liquid–liquid phase chromatography.

The particles are usually made up of hydrated polymers like cellulose or agarose. Such particles can be immobilized and washed with a mobile phase in a column. They have good flow properties and enough mechanical robustness and chemical inertness for biomolecule chromatography. Because biomolecules evolved to function in an aqueous environment, aqueous buffers are typically used as the mobile phase if the molecule's native structure is required (e.g., in the purification of active enzymes).

If the native structure is not necessary, however, additional 'nonbiological' conditions can be used, such as organic solvents (for example, in reverse phase peptide purification). The most prevalent phase systems utilized in biochemistry are liquid–solid and liquid–liquid. Other phases, however, may be employed in specialized situations. In gas chromatography, for example, gas-solid and gas-liquid phases are utilized. Chromatographic separation is a direct outcome of the various K values of each sample component, regardless of the actual phase composition.

4.2.3. Liquid Chromatography

Separations are frequently carried out in aqueous buffers below room temperature to avoid biological activity loss. Low temperatures are particularly significant in chromatography of cell extracts during protein purification, for example. This inhibits the activity of proteases, which would otherwise degrade the protein of interest. Liquid chromatography (LC) is chromatography using liquid mobile phases. LC uses the experimental setup depicted in Figure 4.2. Separation takes place in a stationary phase-containing column. The amount of sample to be separated and the technique of chromatography to be utilized will determine the volume and shape of the column. Buffer is kept in a reservoir and pumped onto the column via tubing. Appropriate valves enable easy sample input into this flow or, if needed, the development of gradients with a second buffer.

The stationary phase is packed in the column, and separation occurs as the sample passes through the bed of the stationary phase. As the material passes through the stationary phase in partition chromatography modes, it separates into discrete components (e.g., gel filtration). However, in adsorption chromatography modes, the complete sample must be loaded first, and then fractionated. Ion exchange chromatography, in which the sample components are eluted by a gradient of competing salt counterions after the sample has been fed onto and washed fully through the column, is a good example of adsorption chromatography. Sample components that have been separated elute from the column and are identified. Figure 4.3 depicts a typical LC separation.

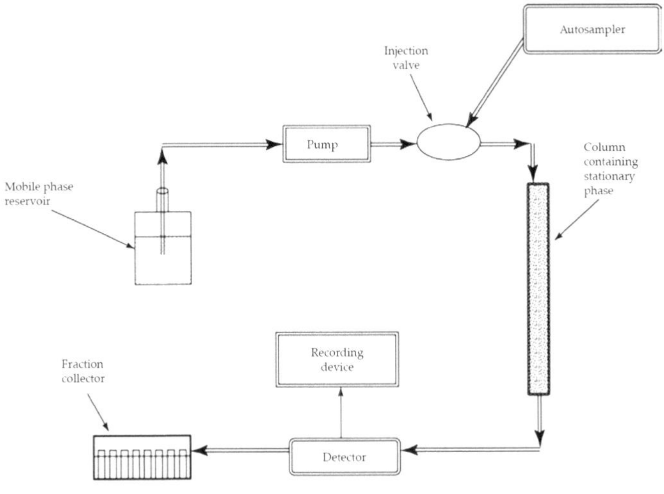

Figure 4.2. A typical liquid chromatography system. Arrows indicate the flow direction. The sample is loaded by injecting it into a valve. If a high number of samples are needed, an autosampler can be utilized to repeatedly refill the column after each chromatography step.

Source: Image by biologicaproceduresonline

4.2.4. Gas Chromatography

Many biomolecules are susceptible to high temperatures, which can cause their structure and function to be destroyed. However, in the temperature range of 200–250 °C, some important biochemical compounds can be transformed into derivatives that are structurally stable but volatile. Trimethylsilated sterols and carbohydrates (esterified at their hydroxyl groups), as well as methylated esters of fatty acids, are good examples.

Few compounds (for example, ethanol) can be stable and volatile at lower temperatures without derivatization. Gas chromatography can be used to examine both of these classes of compounds (GC, also sometimes called gas-liquid chromatography; GLC). This is a type of partition chromatography in which a volatile sample is conveyed in an inert gas mobile phase (carrier gas) such as nitrogen, helium, or argon and applied to a thin column (0.1 to 0.5 cm diameter) containing a liquid stationary phase and ranging in length from one to thirty meters.

A syringe is used to inject the sample into the system through a rubber septum into an injection port. To promote efficient volatilization, this port is kept around 10 degrees Celsius above the column temperature. The column is kept at a very high temperature in an oven. Temperature control in this oven is critical for successful chromatography, and temperature gradients can be utilized to ensure good separation. Because the sample components partition differently between phases and the column is so lengthy, they are effectively separated from one another.

The column can be filled with solid particles covered with a liquid phase or it can be a capillary column with liquid coating on the inner wall of the column. In comparison to packed columns, they have much smaller inner diameters (e.g., 25 μm) and are often longer (10–100 m).

Flame ionization, flame photometry, and thermal conductivity detectors can all be used to detect sample components. The sample is combusted to produce electrons and other ions that form an electric current in flame photometry detectors, whereas the sample alters the conductivity of a wire in flame photometry detectors.

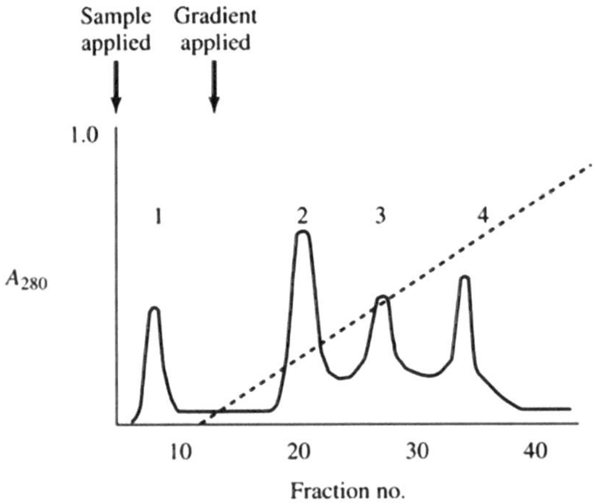

Figure 4.3. A typical liquid chromatography experiment. Ion exchange chromatography was used to separate a group of isoenzymes. With detection at 280 nm, a 0–100 mM NaCl gradient (dashed line) is utilized. Proteins are frequently detected at this wavelength. Peak 1 contains free material, whereas peaks 2–4 contain proteins that are bound with increasing affinity.

Source: Image by biologicaproceduresonline

Detectors are kept around 10 degrees Celsius above the temperature of the column, just like the injection port. The carrier gas flow may be split before it enters the flame ionization detector if the separated sample components are to be collected for additional analysis (or for an interfaced approach such as mass spectrometry). GC separations can be improved by adjusting column temperature, gas flow, and column type.

4.3. PERFORMANCE PARAMETERS USED IN CHROMATOGRAPHY

In biochemistry, several types (e.g., GC, LC) and modalities (e.g., gel filtration, ion exchange) of chromatography are utilized. Comparing and quantifying separations is therefore quite important for optimizing them for a specific purpose. In an analytical HPLC separation, for example, we might want to swap out a column with one from a different manufacturer or use a slightly different stationary phase. The question then becomes, in comparison to the old column, how effective is the new one? Performance criteria are the yardsticks that allow us to conduct such comparisons.

4.3.1. Retention

The various sample components segregate on the column when chromatography is performed as shown in Figures 4.2 and 4.3 because they are retained to different degrees. This retention can be expressed in terms of time (with zero being the time of sample insertion) or volume (calculated from the volume at sample injection). These are referred to as retention time (t_R) and retention volume (V_R) (see Figure 4.4). Equation (4.2) can be used to convert these terms.

$$V_R = f \cdot t_R \qquad \text{Eq (4.2)}$$

Where t_R is the retention period in minutes, V_R is the retention volume (in ml), and f is the flow rate (in ml/min). Each component of the sample will have a characteristic retention (i.e., t_R or V_R) under a single set of experimental circumstances (temperature, column composition and dimensions, etc.) that may be used to identify it in different samples. Retention is inversely related to the flow velocity and proportional to the length of the column used.

Furthermore, because the detector's response is usually proportional to the concentration of the sample components, that may quantify each one by calculating their peak regions.

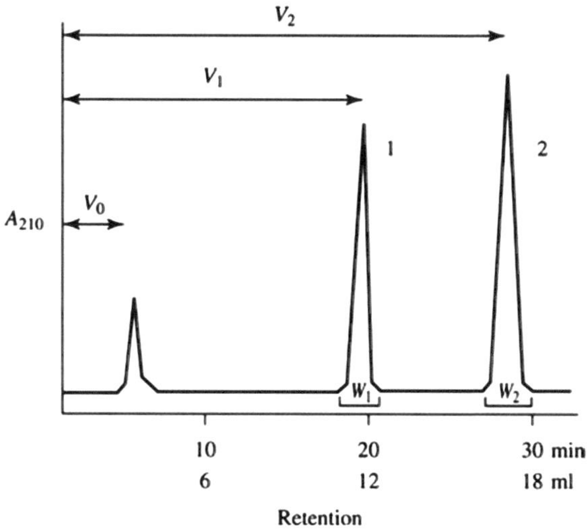

Figure 4.4. Retention in column chromatography. A typical chromatography trace demonstrating the separation of two components (1 and 2). The component retention volumes are shown by V_1 and V_2, respectively, whereas the base peak widths are indicated by W_1 and W_2. V_0 stands for the void volume.

Source: Image by biologicaproceduresonline

The mobile phase moves across the column faster than sample components. As a result, for each component, a retardation factor (R.F.) can be defined as follows:

$$R.F. = \frac{Vsample}{V\ mobile\ phase}$$

Eq (4.3)

V signifies the rate of movement. Even if the sample did not adsorb onto or partition into the stationary phase, it would elute from the column in a finite length of time. The void or excluded volume is the volume required for the injected solvent to simply flow through the column without being held and eluted (V_0, see Figure 4.4). This is the total volume of the column, including the injection valve and tubing, less the volume of the polymeric column packing component. The included volume contains retention volumes bigger than V_0, and it is here that the sample can be fractionated.

4.3.2. Resolution

Figure 4.4 depicts a sample chromatographic separation of two components of a mixture, 1 and 2. We use the term resolution, R, to characterize the separation of the two peaks since they are 'resolved' from each other. This crucial chromatography performance parameter allows us to compare how effective a particular column is at separating a mixture. R has a mathematical definition (Equation (4.4)), which can be deduced from the trace in Figure 4.4:

$R = (V_2 - V_1)/((W_1 + W_2)/2)$ Eq (4.4)

W is the base width of each sample component's peak in the retention volume, V. The best resolution chromatography is those with narrow (i.e., small W) and well-separated peaks (e.g., large V_2 V_1), as shown by this equation.

4.3.3. Physical Basis of Peak Broadening

When a small volume of a single component is given to a chromatographic system, it is likely to elute as a sharp peak of equal volume. However, due to a phenomenon known as band widening, in which the sample elutes in a significantly greater volume than the one in which it was applied, this is rarely observed. This is a crucial factor in determining the R-value of a given chromatography column, according to the factors outlined in Equation (4.4).

Understanding the physical foundation of band broadening is crucial to comprehending why certain chromatography studies accomplish separation while others do not. The applied sample may interact with the stationary phase, resulting in sample diffusion or mass transfer processes inside the mobile or stationary phase. Figure 4.5 depicts these in diagrammatic form.

Eddy diffusion is caused by the mobile phase's proclivity for eddies or localized circular movement, similar to that seen in whirlpools. The mobile phase travels via channels in the stationary phase (which are usually made up of tiny particles) that vary in size, with some being wider and others narrower. The circular flow is often slower in small channels than in wide channels.

The sample will be gradually dispersed between fast-flowing, wide channels and slow-flowing, narrow channels because it will diffuse through a variety of channel diameters. Diffusion causes a band of sample to widen as it travels through the stationary phase.

The distribution of sample molecules within the mobile or stationary phases causes mass transfer processes. Because the mobile phase adjacent to the particle moves more slowly than the mobile phase in the center of the channel, mass transfer happens. This is owing to the particle's frictional resistance to the mobile phase's flow.

Different sample components enter channels that are not oriented in the same direction as the main flow through the column, resulting in stagnant mobile phase mass transfer. In these channels, these molecules are kept in the stationary mobile phase for variable durations of time. As a result, the sample will be kept to varied degrees. In stationary phase mass transfer, mass transfer can also occur within the stationary phase (i.e., for the sample to enter the surface of the stationary phase).

4.3.4. Plate Height Equation

One technique to visualize the chromatographic separation of a sample is to divide the column into very thin, horizontal portions, which we refer to as theoretical plates. It can be clearly seen how separation depends on the difference in component partition coefficients if a basic example is considered of a two-component mixture with partition coefficients of 1 and 0.1, respectively, and see how it behaves as it goes through a small number of successive plates.

The theoretical plate number, N, is an important chromatography column performance measure that is directly connected to the surface area of the particles that make up the stationary phase. It can be computed using the relationship between the retention time, t_R, and width of a chromatographic peak.

$$N = 16.\left(\frac{t_R}{W}\right)^2$$

Eq (4.5)

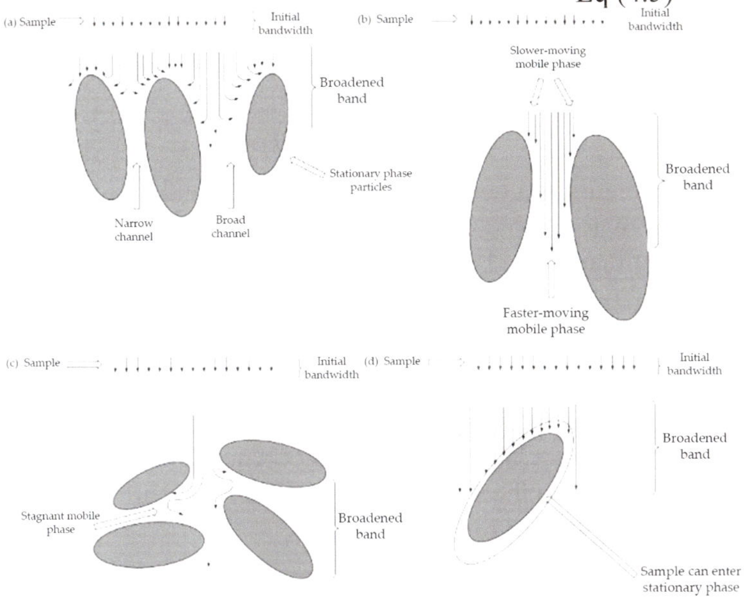

Figure 4.5. Physical causes of band broadening. (a) Eddy diffusion, (b) Mobile phase mass transfer, (c) Stagnant mobile phase mass transfer, (d) Stationary phase mass transfer. All of these factors may work together to broaden the applied sample's comparably limited starting bandwidth.

Source: Image by biologicaproceduresonline

The higher the N number, the better the chromatographic performance of a column can be expected. This value can be used to compare bio separations achieved in different types of experimental settings and even between columns (e.g., electrophoretic separations). In practice, it may be difficult to correctly estimate W, hence an alternative connection using the peak width at half the peak height, W_{ho} (Equation (4.6)) can be used:

$$N = 5.54\left(\frac{t_R}{W_h}\right)^2$$

Eq (4.6)

Two chromatographic columns of differing lengths (for example, 15 and 30 cm) may contain the same number of theoretical plates (for example, a normal 15 0.46 cm C-8 reverse phase column holds around 13000 theoretical plates). The shorter column would obviously be considered to have greater chromatographic performance. The plate height number, H, also known as the height equivalent to a theoretical plate, HETP (Equation (4.7)), can be used to express the number of theoretical plates as a function of column length:

$$H = L/N \qquad \text{Eq (4.7)}$$

where L is the length of a column. This number should be lower the better a column's chromatographic performance. The H value of the former (1.15 µm) will be half that of the latter (2.3 µm) in our example of 15 and 30 cm long columns.

Because W is the numerator in Equation (4.5), the number of theoretical plates in a chromatography column and the level of band broadening experienced by the sample as it passes through the column are closely related. A column with a large number of theoretical plates should generate little band broadening in the sample, whereas columns with a small number of theoretical plates should cause the opposite.

The class of stationary phase used, the particle diameter d_p, the flow rate v, and the molecular size of the sample molecules are the main parameters that lead to substantial peak widening and thus large values for H. (which will affect D_m). By selecting the right stationary phase particle and, in particular, using particles with tiny diameters, the chromatographer can get lower H values. This is the physical foundation for excellent resolution in HPLC and FPLC systems. It's also worth noting that using slower flow rates and a higher temperature can potentially improve separations (which tends to increase D_m).

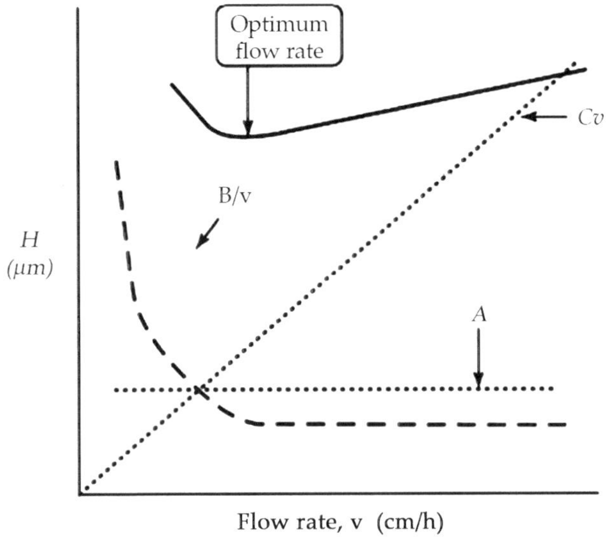

Figure 4.6. The van Deemter curve. A plot of H versus flow rate, v, is shown (solid line).

Source: Image by biologicaproceduresonline

4.3.5. Peak Symmetry

Many chromatographic separations accomplish baseline separation, in which individual peaks in the chromatogram do not overlap. Furthermore, the peaks are frequently believed to have a Gaussian form. This is the optimum circumstance, and Equations (4.5) and (4.6) may allow us to determine N straight from the peak. Peak symmetry can be greatly skewed from Gaussian in practice.

This is due to a number of reasons, including the use of nonlinear flow rates, poor peak separation (e.g., the presence of 'shoulders' on peaks, which is a key indicator of peak contamination), and the use of gradient elution techniques, all of which are common in practical chromatography. As a result, peak symmetry can provide useful information about the performance of chromatography systems.

4.3.6. Significance of Performance Criteria in Chromatography

Performance criteria can be used to compare chromatography systems that are otherwise quite dissimilar. Because they rely solely on physical interactions between sample components and the stationary phase, they are unaffected by the chemical composition of sample adsorption to the stationary phase.

As a result, these criteria can be used to compare reverse phase HPLC separations to open column ion-exchange chromatography with DEAE-cellulose, for example. This comparison is also independent of the sample components' nature, therefore it can be used to compare peptides and sterol derivatives, for example. High resolution (R), a large number of theoretical plates (N) in the column, a low plate height number (H), and high values of the separation factor (α) are all required for 'excellent' chromatography. These criteria can easily identify poor chromatography, and their application can be used to quantify better purification.

4.4. CHROMATOGRAPHY EQUIPMENT

4.4.1. Outline of Standard System Used

To obtain a chromatographic separation, a variety of experimental systems can be used. The degree of automation utilized determines the system's level of complexity. The general arrangement, however, is as indicated in Figure 4.2. In the laboratory, the simplest manual chromatography set-up for open column chromatography can be created utilizing plastic bottles as reservoirs and glass columns blown over sintered glass funnels. After separation, manual absorbance measurements can be used to detect in such a system. Flow in such a system can be driven either by gravity or by a peristaltic pump. Commercial providers offer a wide range of low-cost glass and plastic columns.

The main benefit of automation is the ability to achieve more repeatable separations. Peristaltic pumps, on-line detectors, and fraction collectors can provide a minor level of automation. However, the introduction of specialized microprocessor-controlled chromatography workstations has enabled sequential separation of multiple samples, fine-tuning of chromatography programs to optimize resolution, automatic peak collection and calculation of peak area and symmetry, and chromatogram overlaying for comparison.

4.4.2. Components of Chromatography System

To remove particle matter and other pollutants from buffers and other solutions used in the preparation of the mobile phase, they should be filtered well before use. They should also be degassed for high-resolution chromatography, while some systems have automatic sparging with oxygen-free nitrogen or helium gas. The mobile phase is pumped onto the column from a reservoir. When more than one reservoir contributes to the mobile phase (as in binary, ternary, or quaternary gradients), the flow from these reservoirs is combined in a mixer before reaching the column.

Flow in open column chromatography can be achieved by gravity or peristaltic pumps. High-precision pumps propel the mobile phase through the system in FPLC and HPLC. Separation takes place in a stationary phase-containing column. Depending on the system's pressure, this could be made of glass, plastic, or steel. In high-performance liquid chromatography, a guard column may be used before the chromatography column in the flow. This serves to keep the later column from becoming clogged and to extend its useful life.

Absorbance, fluorescence, refractive index, electrochemical, or other physicochemical measurements may be used for on-line detection. Most proteins may be detected at 280 nm, but to avoid missing proteins or peptides with few aromatic residues, detection in the range 205–220 nm (which detects peptide bonds with λmax at 214 nm) is generally preferred. Greater sensitivity is possible with fluorescence and refractive index detection, which is ideal for analytical chromatography. An elution profile is a graph of detector signal vs. sample component retention. A fraction collector can be used to collect separated peaks.

Prior to chromatography, the sample should be filtered or centrifuged to remove any particles that could clog the column. In open column chromatography, samples are introduced directly onto the column, while in high-resolution systems, samples are injected into the mobile phase stream through a sample injection port. An autosampler is used for numerous samples, which is very beneficial in analytical separations. Because open column separations are often sluggish, they are often performed at lower temperatures, such as in a cold room or a refrigerated cabinet, to reduce biological activity loss. At room temperature, higher resolution chromatography is usually possible.

4.4.3. Stationary Phases Used

Hydrated polymers are the most typically employed in biochemical sample chromatography. Commercially accessible are a variety of these and other packings, as well as a selection of these, along with their applications. Under the chromatographic conditions utilized, the stationary phase must be mechanically stable (particularly to high pressures and liquid shear forces) and chemically inert. It is frequently desired that it can operate as a support for chemical groups with favorable chromatographic properties such as hydrophobicity, ion exchange, or affinity ligands.

4.4.4. Elution

Depending on the chemical interaction between sample components and the stationary phase, elution from or development of chromatography columns can be accomplished in one of three methods. Figure 4.7 depicts these elements. The mobile phase flowing through the column is continuous in both flow rate and composition in continuous flow elution, also known as isocratic elution.

This is especially true in systems that use partition chromatography. With this sort of elution, it is common to see that leading (i.e., early eluting) peaks are sharp but poorly resolved, whereas trailing (i.e., late eluting) peaks are well separated but have undergone considerable peak broadening. This is referred to as the general elution issue.

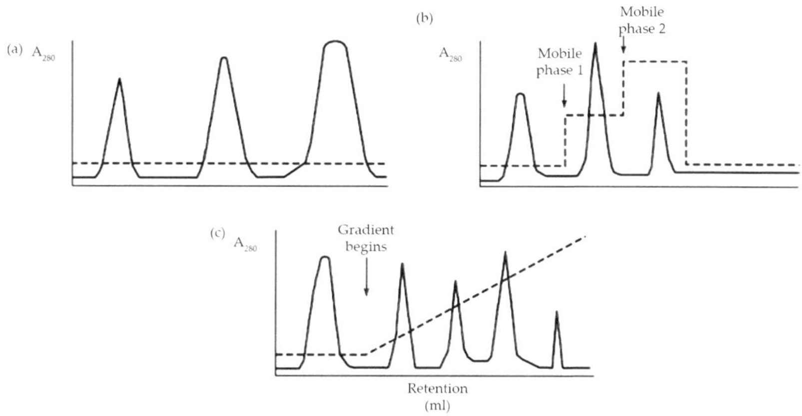

Figure 4.7. Elution from stationary phase. The mobile phase is represented by dashed lines. As explained in the book, the composition of this may change to give various pH, ion strength, or hydrophobicity. (a) Continuous flow elution. When sample components have distinct inherent affinities for the stationary phase and the mobile phase composition and flow rate remain constant, the sample components separate. (b) Batch flow elution. Adsorbed sample components can be eluted selectively by washing the column with a variety of mobile phases in a sequential manner. (c) Gradient elution. Adsorbed material is separated using a gradient consisting of two or more buffers with a continuously

changing mobile phase composition. Continuously adjusting % buffer B in the mobile phase can produce more complex gradients (e.g., hyperbolic gradients).

Source: Image by biologicaproceduresonline

When sample components have adsorbed to the column packing, their affinities for the stationary phase are sometimes considerably different. A description of the major types of adsorption chromatography used in biochemistry; it will be obvious from this that the different affinities derive from the sample components' underlying structural properties.

This could be used by the chromatographer to separate them. Batch flow elution, which involves the progressive introduction of multiple mobile phases in which the pH, polarity, salt content, or any other component is altered, can be employed in this case. One sample component may be bound to the stationary phase while another elutes, usually in a limited amount, under certain conditions.

Gradient elution is a variation on this, in which the pH, polarity, and salt concentration are adjusted continuously rather than gradually. This method of elution overcomes the previously described general elution problem by ensuring high-resolution separations across the chromatogram within a relatively limited elution range. This process necessitates the use of a gradient maker. The development of complex gradients with a range of shapes is now possible thanks to automation (Figure 4.7).

Gradients are made by continually mixing two separate buffers, commonly referred to as buffers 'A' and 'B', respectively. Ternary gradients, made from three different buffers, are also feasible. The mobile phase is formed as a result of this mixing. To calculate elution conditions for batch flow elution, gradient elution would be required. This method is frequently employed when scaling up chromatography for preparative separation, such as in protein downstream processing. Combining several gradient steps with isocratic elution may improve the separation of poorly resolved peaks in some circumstances.

In ion exchange chromatography, elution with ionic substances like NaCl and KCl may necessitate high salt concentrations. After chromatography, this salt will need to be eliminated in downstream protein processing, which can be costly and time-consuming. Stepwise elution with displacers, or molecules having a higher affinity for the stationary phase than the target protein of interest, is another option. These molecules can have a low or high molecular weight (for example, heparin and starch derivatives) and have the benefit of inducing protein elution in highly concentrated peaks of pure material at low displacer molar concentrations (e.g., 40 mM).

Because biomolecules like proteins are so sensitive to their surroundings, changing the pH, ion strength, and other variables in the mobile phase can change the charge, hydrophobicity, and overall structure of their surfaces. As a result, if the pH of the mobile phase is changed, the elution profile achieved at that pH will change. Changes in pH will impact proteins differently because they are highly individual in their sensitivity to their environment. Thus, we can optimize separation conditions by performing chromatography with a variety of stationary phases under a variety of pH and ion strength settings.

4.5. MODES OF CHROMATOGRAPHY

So far, there have been discussions about chromatography systems in terms of resolution and performance. This is useful because it explains why the physical properties (particularly particle diameter) of the particles that make up the stationary phase are so crucial. The chemical basis underpinning the interaction between sample components and the stationary phase, on the other hand, provides a complementary description of the major biochemistry chromatography systems. The mode of chromatography can be described as a result of this interaction.

Regardless of the physical features of their stationary phases, all of these systems are available in a number of chromatographic modes. The matrix depicted in Figure 4.8 exemplifies this circumstance. Only two criteria, chromatography mode and format, can be used to locate any chromatography system on this matrix (i.e., stationary phase differing in physical characteristics such as particle diameter). This matrix is valuable since it can be quickly expanded to support additional chromatographic formats and modes.

4.5.1. Ion Exchange

Proteins contain a net charge on their surface that is primarily derived from amino acid side chains, which can be positive or negative depending on the pH. Acidic amino acids (glutamic and aspartic acids) produce negative charges around pH 7.0, whereas basic amino acids (lysine and arginine) contribute positive charges. Most proteins have a net positive charge at acidic pH and a net negative charge at alkaline pH. Positive and negative charges can cancel each other out, resulting in a protein with net zero charge at a given pH value, even if it includes some positive and negative charges. The isoelectric point, pI, is the pH at which a specific protein has no net charge (Figure 4.9).

MODE →

FORMAT ↓

	Ion exchange	Gel filtration	Reversed phase	Hydrophobic interaction	Affinity	IMAC
Open column	DEAE-cellulose CM-cellulose	Sephadex G-25		Octyl-sepharose Phenyl sepharose	Glutathione -agarose	IDA-agarose
FPLC	Mono-P Mono-Q	Superose-6 Superose-12	C_8-silica C_{18}-silica	Phenyl-sepharose	NHS-activated Superose-12 Protein A/G-Superose	IDA-Superose
HPLC	DEAE/CM -silica	Fractogel TSK	C_{18}-silane	Propyl-hydroxylated polyether		IDA-silica
Perfusion	Poros-HQ Poros-HS		Poros R_1 Poros R_2	Poros-PE Poros-ET	Poros-A (protein A)	Poros-IDA
Membrane	DEAE/CM -cellulose				Activated -cellulose	IDA-cellulose

Figure 4.8. The link between different chromatography systems in terms of chromatography mode and format is depicted in this matrix. A selection of regularly used stationary phases is depicted.

Source: Image by biologicaproceduresonline

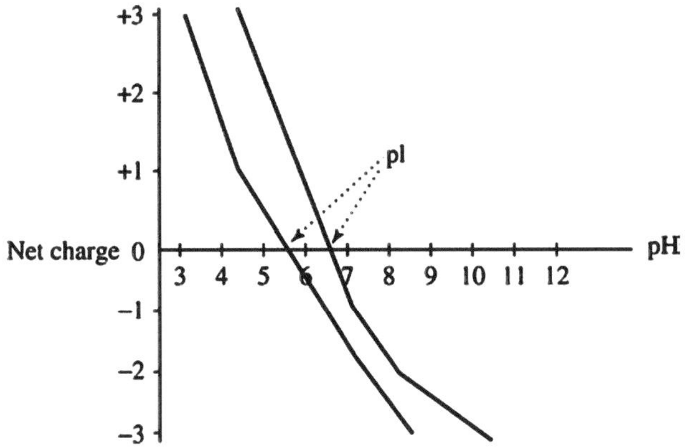

Figure 4.9. Surface charge of proteins. At a variety of pH values, the net charge on the surface of two proteins (pI values of 5.5 and 7.5, respectively) is depicted. Take note of how this changes with pH.

Source: Image by biologicaproceduresonline

Because surface charge, structure, and pI are all determined by the amino acid sequence of a protein, they can be used to distinguish one protein from another.

Ion exchange chromatography is a type of adsorption chromatography in which charged proteins or other biomolecules are exchanged for tiny ions with similar charges found in salts (e.g., Na+, Cl). These ions are attached to ion exchange groups or ion exchangers, which are chemical structures on the surface of the stationary phase.

In biochemistry, there are two types of ion exchangers: anion exchangers (which exchange negatively charged ions or anions) and cation exchangers (which interchange positively charged ions or cations) (which exchange positively charged ions or cations; Figure 4.10). It's vital to note that this exchange mechanism is unaffected by the molecule's mass, shape, or other physical features. On ion exchange chromatography, proteins that are otherwise quite diverse could have the same behavior: Isoenzymes, on the other hand, may behave considerably differently despite having the same fundamental structure and biological activity.

Ion exchangers are attached to cellulose, agarose, and vinylbenzene, among other solid supports. The physical parameters of the stationary phase as a whole (ion exchanger plus support) will decide whether a specific experimental setup can achieve high or poor resolution. Some ion exchangers are weak, meaning they work best across a restricted pH range, while others are strong, meaning they work well throughout a wider pH range. Strong anion and cation exchangers are those with quaternary amino or sulfonic acid groups, while weak anion and cation exchangers are those with aromatic/aliphatic amino and carboxylic acid groups. The pI of the protein of interest has a big role in deciding which ion exchange stationary phase to use.

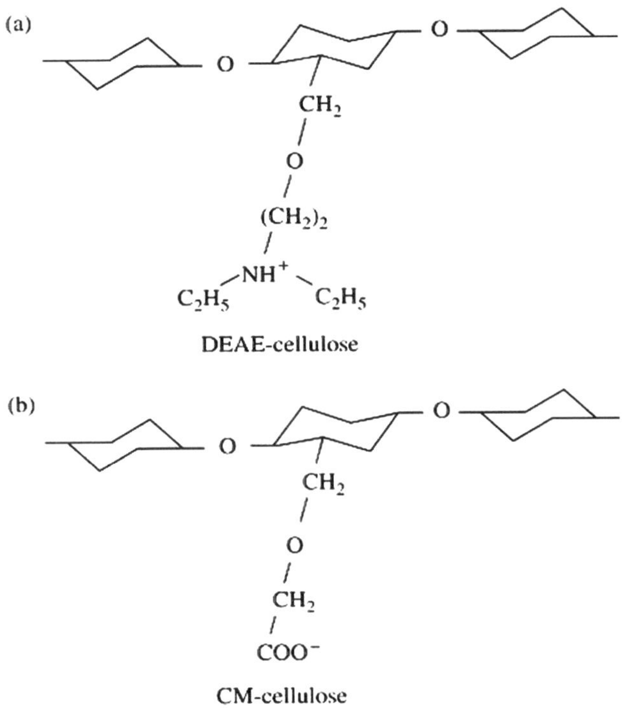

Figure 4.10. Ion exchangers. The structure of (a) diethylaminoethyl (DEAE)-cellulose, an anion exchanger and (b) carboxymethyl (CM)-cellulose, a cation exchanger.

Source: Image by biologicaproceduresonline

The sample is put to a stationary phase that has been charged with the ion to be exchanged; the counterion (e.g., Cl) in an ion exchange chromatography experiment. The protein (and any other species in the sample with a similar charge) may bind to the ion exchanger by exchanging with this counterion. It is critical that the sample be free of components of similar charge, as these are often present in high molar excess relative to the protein content.

The ion exchanger will not distinguish between proteins with a single positive charge and other ions in the sample, such as Na+. Before ion exchange, desalting is done using gel filtration chromatography, dialysis, or centrifugal filtration.

Elution of bound proteins is accomplished by reversing the binding process and substituting a counterion for the protein once more. This is commonly accomplished by using a significant amount of a specific salt containing the counterion in the mobile phase (e.g., NaCl). Because proteins have varying net charges, they can bind to an ion exchanger with varying degrees of intensity at a given pH. For example, certain proteins may bind strongly while others bind weakly or not at all. By administering salt in a continuous gradient, we can use this to separate proteins. In such a system, weakly bound proteins will elute first, followed by highly bound proteins.

At all pH levels, proteins have some charge (their net charge is zero only at the pH corresponding to their pI). We can change the net charge on their surface selectively by adjusting pH. Even if two proteins behave the same at one pH, they are unlikely to behave the same at all pH values. We can resolve proteins that might chromatograph at a single pH by performing separations throughout a pH range.

It's important to keep in mind that many proteins are physically unstable at pH extremes while looking at chromatographic behaviors throughout a variety of pH values. This may cause the stationary phase bed to foul, as well as a loss of biological activity in the purified protein. Prior to ion exchange chromatography, preliminary tests to evaluate pH stability and solubility of the protein to be purified are recommended.

Ion exchange chromatography can be used to separate fully active proteins under native conditions, but it can also be used to separate them under denaturing conditions. In the presence of 4–7 M urea, peptides from a chemical or enzymatic digest of a protein, for example, can be isolated.

Urea is a chaotropic substance that denatures proteins by breaking ionic connections (particularly hydrogen bonding) within and between protein subunits or individual peptides in a digest and diminishing hydrophobic interactions within polypeptides. Subunits and peptides are no longer held together by interactions such as salt bridges and hydrogen bonds when ion exchange is undertaken in the presence of urea. Despite being a polar molecule, urea has no net charge and hence will not interact with the ion exchanger.

4.6. HIGH-PERFORMANCE LIQUID CHROMATOGRAPHY (HPLC)

4.6.1. Equipment Used

Stagnant mobile phase mass transfer, H3, is a significant component of the overall plate height equation. This is the diffusion of sample components into and out of the stationary phase's stagnant mobile phase. The square of the particle diameter is directly proportional to this term (d^2_p). As a result, reduced d_p values are projected to result in huge decreases in H and significantly enhanced chromatographic performance. One physical reason for this is that the surface area available for adsorption/desorption is proportionately much bigger in such particles than in larger particles.

This is why the theoretical plate number, N, is proportional to the stationary phase particle's surface area. As a result, adsorption/desorption kinetics and, as a result, chromatographic performance should be greatly improved in stationary phases with low d_p values.

Smaller diameter particles, on the other hand, result in significantly lower flow rates because they provide higher resistance to mobile phase flow. This causes back pressure on the stationary phase, which can result in substantial band broadening and extended retention durations due to the stationary phase bed being distorted. Inversely proportional to d^2_p, back pressure rises. The stationary phase particles employed in open column chromatography are mostly polysaccharide gels, which are mechanically weak.

As a result, even if such particles had small diameters, they would not be robust enough to endure the high pressures required for high-resolution chromatography. High flow rates can be attained by aggressively pushing the mobile phase through the stationary phase under high pressure with silica-based particles (diameter 5–10 μm) (up to 55 MPa).

This is referred to as high-performance liquid chromatography (HPLC). In this type of chromatography, a combination of high pressure and small particle diameter achieves great resolution. The high pressure required by the system dictates the equipment's design and construction. In reality, the pressure required for flow in HPLC systems becomes unfeasible below d_p values of 3–5 m.

The sample is injected using a syringe that can be operated manually or mechanically from an autosampler. The septum is made of neoprene and

Teflon. This latter approach is particularly useful for analytical applications, as it is particularly suited to situations in which many, repeated analyses may be required.

It is essential for the flow of the mobile phase to be extremely stable in order to accomplish the goal of achieving a stable baseline in the elution profiles (i.e., pulse-free). In high-performance liquid chromatography (HPLC), the pulse-free flow of the mobile phase can be accomplished with either constant volume (also known as constant displacement) or constant pressure pumps. During chromatography, the latter keeps the pressure at a constant level regardless of any changes in the resistance of the stationary phase to flow. On the other hand, if this resistance were to grow, then the flow rate would automatically drop in order to keep the pressure constant. Because they keep a constant flow rate through the stationary phase throughout chromatography and can increase pressure as needed to offset increased resistance from the stationary phase, constant volume pumps are the most common type of pump used in high-performance liquid chromatography (HPLC).

The reciprocating pump is the most common type of constant volume pump employed. This type of pump employs a piston to transfer a predetermined amount of solvent onto the column in a series of repeated cycles of filling and emptying. Pulses may be introduced into the flow by these pumps; however, pulse dampeners are integrated into them to reduce the likelihood of this happening.

The mobile phase in most types of high-performance liquid chromatography (HPLC) is a mixture of two or more components that is applied to the column in order to produce quick separation. As a result, the HPLC instrument comes equipped with a mixer for more effective mixing.

In HPLC, a wide number of different methods of detection are utilized. In the study of biological materials, the use of variable-wavelength UV detection has become increasingly common. It is possible to carry out detection at 280 nm for proteins, which is specific for aromatic amino acids, or at 220 nm, which is specific for peptide bonds; however, 260 nm is useful for the detection of nucleic acids. The fluorescence detector and the refractive index detector are two other types of detectors.

Postcolumn derivatization is an example of a specific method of detection that can be utilized, for instance, in amino acid analysis. This method was developed in the 1970s. Chromatography is used to separate the amino acids, and after that, the amino acids are derivatized using o-phthalaldehyde (OPA) postcolumn. The OPA conjugates that are produced are highly fluorescent and can be identified with a high degree of sensitivity. There is a vast range of possible dimensions for HPLC columns.

Flow rates of 1–10 ml/min are possible using a typical column with an inner diameter of 2–5 mm. However, finer columns are being employed in high-resolution separations, resulting in a drop in the flow rates that can be achieved: Inner diameters of capillary columns range from 100 μm to 1 mm (flow rate range: 0.4–200 μl/min); nanobore columns have inner diameters of 25–100 μM (flow rate range: 25–4000 nl min).

Nanobore HPLC yields extremely steady submicroliter flowrates, making it easier to interface with other analytical methods like mass spectrometry. These are very significant and beneficial in situations like proteomics, where only little amounts of protein are available for analysis.

4.6.2. Liquid Phases in HPLC

It is absolutely necessary for the mobile phase that is used in HPLC to have a high level of purity and to be chemically inert. When air or gas is mixed with solvents, bubbles can occur at elevated pressure. These bubbles have the potential to damage stationary phase packing and interfere with the chromatographic analysis. Before performing chromatography, it is customarily essential to degas the solvents, which can be done either by purging the solvent with gas, which is also frequently termed sparging, or by subjecting the solvent to a vacuum.

In order to prevent the fouling or blocking of columns caused by particulate matter that is suspended in the sample and the solvents that are used, it is standard practice to pass both of these things through a filter with a pore size of 5 microns and to place a guard column in the system before the chromatography column.

In high-performance liquid chromatography (HPLC), different mobile phase solvents can be utilized. The selection of these is frequently dependent on the results of experiments; nevertheless, they can be arranged into a series, known as an elutropic series, that reflects how well they displace components of the sample that have been adsorbed.

In general, an increase in polarity is accompanied by an increase in the value of the eluent's strength. When selecting a mobile phase solvent, it is essential to make sure that the solvent does not impede the detection of any of the sample's components in any way. Strong absorbance of ultraviolet light, which must be accounted for in any analysis using this method of detection.

4.6.3. Two-Dimensional HPLC

It is not always possible to obtain a complete separation of a biochemical extract using just one chromatogram. This is especially true for larger extracts. This is due to the possibility that there is a close similarity between components in terms of polarity, molecule mass, charge, pI, and a variety of other characteristics. However, in order to attain exceptionally high resolution, one must first separate the molecules based on one physicochemical parameter (such as charge characteristics), and then separate the molecules based on a second parameter (such as molecular mass).

This is the fundamental idea of two-dimensional high-performance liquid chromatography (2-D HPLC). These types of multidimensional separations have a very high peak capacity, which refers to the total number of peaks that are distinguished theoretically. For example:

$$P_{MD} = P_1 \times P_2 \times P_3 \ldots \qquad \text{Eq (4.8)}$$

Where PMD, P1, P2, and so on represent the peak capabilities of multidimensional, one dimension, second dimension, and so on separations, respectively. Ion exchange coupled with reversed phase and gel filtration coupled with capillary electrophoresis are two examples of common dimension combinations. When a fraction elutes from the first dimension, it is immediately transferred to the second dimension.

It is possible to achieve separations of exceedingly complicated combinations of peptides or proteins with this method due to the exceptionally high resolution achievable. In addition to connected columns, it is also feasible to have single columns that include two different forms of packing. An ion exchanger that has been combined with a stationary phase that has its phase reversed is one example. The use of two-dimensional high-performance liquid chromatography (2-D HPLC) as a supplement to two-dimensional electrophoresis holds promise in the field of proteomics.

4.7. MEMBRANE-BASED CHROMATOGRAPHY SYSTEMS

4.7.1. Theoretical Basis

Systems of chromatography that make use of stationary phases that are composed of particles of a diameter, denoted by d_p. However, low dp values necessitate the use of high-pressure chromatography, which is a must give the significance of this parameter in the process of generating high resolution separations. In addition, the d_p has a lower limit set between 3 and 5 μm as a consequence of the impractically high back pressures that occur below this figure.

Utilizing stationary phases that are based on membranes, as opposed to those that are based on particles, is one alternate method. The stationary phase of membrane-based chromatography is made up of layers of thin, porous membranes that are placed on top of one other. In order to provide a variety of chromatographic modes, they can be coated with ion exchange, affinity, or other groups. The membranes are housed within a cartridge, which can be substituted for a column in open column, FPLC, and HPLC formats respectively.

It has been discovered that diffusion does not place any limits on the amount of mass that can be transferred to the stationary phase when a sample is introduced into such a system. This process seems to occur predominantly by convection, resulting in substantially quicker adsorption kinetics (for example, two hundred to three hundred times faster than in agarose-based systems).

In addition, it is feasible to reach significant capacities in the cartridge (15–30 mg of protein) by layering numerous membranes on top of each other (to a thickness of 10 mm) and by increasing the width of the membrane. It is also possible to employ extremely high flow rates (up to 10 ml/min when operating under normal conditions) without producing significant amounts of back pressure.

Large transport pores contribute to a high surface area available for adsorption within the membrane, while a broad pore size distribution in the membrane appears to be the cause of the unusual observations that have been made. This distribution of pore sizes is responsible for the high flow rates that have been observed. In these regards, the method is analogous to the chromatographic technique known as (particle-based) perfusion chromatography.

4.7.2. Applications of Membrane-Based Separations

Cartridges designed specifically for membrane-based chromatography can be purchased from a number of different vendors. These cover the most important chromatographic techniques, with the exception of gel filtration chromatography. Because antibodies, metals, and several other ligands are essentially fixed to the membrane during affinity applications, membrane chromatography is one of the most common applications of this technique.

It appears that membrane-based affinity chromatography provides a substantially higher efficiency of ligand consumption than agarose-based systems do, which is one of the advantages of using this method. This is essentially the result of a higher degree of convection, which enables a greater degree of access to immobilized ligands. Another benefit of membrane-based systems is that, in comparison to column-based systems, they are not as constrained in terms of the geometry of their designs.

In contrast to chromatography columns, filtration systems can be constructed in a very wide variety of different ways. It is possible to imagine downstream processing, for example, in a membrane-based system that would not be possible in a column due to the limitations imposed by the dimensions of the column, the properties of the particles, and the effects those properties have on flow rate, back-pressure, and resolution. This is something that is imaginable but not possible. The use of cross-flow filtration is one illustration of the unconventional design formats that are at your disposal.

4.8. CHROMATOGRAPHY OF A SAMPLE PROTEIN

4.8.1. Designing a Purification Protocol

The sequence of operations that must be carried out in order to remove impurities from a protein is known as the "purification protocol." When purifying proteins, it is often necessary to strike a delicate balance between a number of variables that may be interconnected in a highly convoluted fashion. The following are the most significant of these:

Stability characteristics of the protein being targeted:

When it comes to their resistance to changes in temperature, pH, and oxidation, proteins have a wide range of characteristics that set them apart from one another. The protein may denature either when it is withdrawn from the cell or when it is expressed in the foreign environment of a heterologous expression system. This results in the loss of the protein's structural integrity and the biological activity that is associated with it. It is possible that this does not make a difference if only a tiny amount of protein is needed for structural research such as protein sequencing. In most cases, however, a protein that is both physically and functionally intact is necessary. During the purifying process, this sensitivity of the proteins might be utilized to one's advantage.

It is possible to use heat and pH treatments selectively to denature contaminating proteins without damaging the protein of interest when the protein displays remarkable stability to heat and pH. Inadvertent release of proteases from subcellular organelles during the production of cell extracts is a difficulty related to cell disruption.

This can result in the proteolytic degradation of the protein of interest. For example, the release of cathepsins from lysosomes. Included in the buffers used for homogenization and chromatography are protease inhibitors, which can stop this from happening. Incorporating the metal chelator ethylenediaminetetraacetic acid (EDTA), for instance, is one method of reducing the amount of metalloproteinase activity.

Concentration:

The concentration of proteins found inside the cell is of the order of 200 milligrams per milliliter. When we extract proteins from cells, the protein concentration is diluted (typically to less than 1 mg/ml), which can have an impact on the structural stability of the proteins. Because protein purification is a process that involves maintaining the activity of interest while simultaneously reducing the concentration of protein, this can mean that biological activity may decrease while the protein is being purified.

From one stage to the next, we see a declining specific activity, which is expressed as activity per mg of protein. The presence of a stabilizer in the extract, such as bovine serum albumin, which can keep the protein content high enough to continue activity, has the potential to reduce this effect to a significant degree.

Expression system:

When employing heterologous expression systems, it is possible to acquire expression levels that fall anywhere between 5 and 20 percent of the total cell protein. It's possible that the cell will become damaged as a result of this, and it's also possible that the cell won't have the enzymes or other proteins it needs to perform post-translational processing correctly.

Because of this, the protein could be found in the body in the form of insoluble plaques or inclusion bodies (also sometimes called refractile bodies). Because they have a density that is significantly higher than that of the other proteins in the cell, they may be isolated from the other proteins by using centrifugation. After then, the plaques can be dissolved by adding urea to the mixture, and the protein can be refolded in the correct manner.

The purification process can be modified to account for the incorrect expression of protein in plaques by this method. Heterologous expression systems provide an additional opportunity, which is the capability of expressing the protein as a fusion that includes polypeptides that are able to recognize affinity-ligands such as glutathione or immobilized materials. This is one of the options available. The incorporation of a protease cleavage site makes it possible to eliminate the affinity site from the purified fusion protein at a later time.

4.9. CONCLUSION

Chromatography as a separation method has a great number of advantages over older techniques such as crystallization, solvent extraction, and distillation. It is capable of separating all the components of a multicomponent chemical mixture without any need of extensive foreknowledge of the identity, number, or relative amounts of the substances present.

It is versatile in nature, that is, it can deal with molecular species ranging in size from viruses, which are composed of millions of atoms to the smallest of all molecules such as hydrogen, which contains only two. In addition to this, it can be used with small or large amounts of material. Some types of chromatography can detect substances present at the attogram (10−18 gram) level, and therefore making the method a superb trace analytical technique extensively used in the detection of chlorinated pesticides in biological materials and the environment, in forensic science, and in the detection of both therapeutic and abused drugs. Its resolving power is unequaled among separation methods.

REFERENCES

1. Sheehan, D., 2009. *PHYSICAL BIOCHEMISTRY: PRINCIPLES AND APPLICATIONS*. 2nd ed. [ebook] Ireland: A John Wiley & Sons, Ltd, Publication, p.5. Available at: <http://biochem.du.ac.in/web/uploads/9%20PHYSICAL_BIOCHEMISTRY_David_Sheehan.pdf> [Accessed 14 June 2022].
2. Jacobson, K., 1962. Ribonucleotides of RNA: Separation by Chromatography on Sheets of Diethylaminoethylcellulose. *Science*, [online] 138(3539), pp.515-516. Available at: <https://www.science.org/doi/10.1126/science.138.3539.515> [Accessed 4 July 2022].
3. Macrae, R., 1986. High performance liquid chromatography in biochemistry. *Food Chemistry*, [online] 21(3), pp.239-240. Available at: <https://linkinghub.elsevier.com/retrieve/pii/0308814686900257> [Accessed 4 July 2022].
4. Simoni, R., Hill, R. and Vaughan, M., 2002. The Use of Chromatography in Biochemistry. *Journal of Biological Chemistry*, [online] 277(40), p.e1. Available at: <https://www.jbc.org/article/S0021-9258(18)36531-1/fulltext> [Accessed 4 July 2022].
5. Simoni, R., Hill, R. and Vaughan, M., 2002. The Use of Chromatography in Biochemistry. *Journal of Biological Chemistry*, [online] 277(40), p.e1. Available at: <https://linkinghub.elsevier.com/retrieve/pii/S0021925818365311> [Accessed 4 July 2022].

CHAPTER 5

MASS SPECTROMETRY AND SPECTROSCOPIC TECHNIQUES

CONTENTS

A method of analysis called mass spectrometry (MS) determines the mass-to-charge ratio of charged particles. The elemental makeup of a sample or molecule can be determined, as well as the masses of the particles. According to the MS principle, chemical compounds are ionized to produce charged molecules or molecule fragments, and then their mass-to-charge ratios are measured using a number of procedures (such as EI/CI/ESI/APCI/MALDI).

5.1. INTRODUCTION TO MASS SPECTROMETRY

Mass spectrometry is an effective method for discovering unknowns, researching the molecular structure, and exploring the basic tenets of chemistry. The identification and quantification of pesticides in water samples, and the identification of anabolic steroids in athletes are examples.

The detection of metals in water samples at PPQ (Parts Per Quadrillion) levels, the carbon-14 dating of the Shroud of Turin using only 40 mg of sample, the search for life on Mars, the determination of the mass of a 28Si atom with an accuracy of 70 ppt, and the analysis of the impact of molecular collision angle on reaction mechanisms are all applications of mass spectrometers.

In essence, mass spectrometry is a method for "weighing" molecules. Obviously, a traditional balance or scales are not used for this. The movement of a charged particle called an ion in an electric or magnetic field provides the basis for mass spectrometry instead.

This motion is influenced by the ion's mass-to-charge ratio (m/z). The mass to charge ratio serves as a measurement of an ion's mass because the charge of an electron is known. The chemistry of ions, the production of gas phase ions, and mass spectrometry applications are the main topics of mass spectrometry study in general.

The earliest types of mass spectrometry date back to Goldstein's and Wien's observations of canal rays in 1886 and 1899, respectively. The route of the cathode rays (electrons) was bent during Thompson's subsequent discovery of the electron in order to ascertain their charge-to-mass ratio. Aston made the first isotope measurements later, in 1928.

Since then, there has been a lot of activity [regarding both new types of ion separators as well as minor advancements in mass spectrometer components, such as various types of instrument interfaces (direct injection, GC, and HPLC)] to various ionization sources (electron and chemical ionization).

For instance, time of flight mass filters were created by Stephens in 1946, ion cyclotron resonance mass filters by Hipple and Thomas in 1949, quadrupole mass filters by Steinwedel in 1953, and double focusing magnetic sector mass filters by Mattauch and Herzog in the 1960s.

Golhke et al. introduced mass spectrometry to GC in 1956 as a technique of sample introduction, and in the middle of the 1980s, electro-spray ionization was used to connect it to HPLC (Blakely and Vestal, 1983; Yamashita and Fenn, 1984). Even as recently as 1985, Hillenkamp and Michael Karas created the MALDI technique (a laser-based sample introduction device), which significantly improved the analysis of protein structures.

More varieties of mass analyzers will undoubtedly be created. This chapter will only cover the fundamental mass spectrometer equipment that is used to analyze organic chemicals that are released from GC and HPLC systems.

It also applies to the effluents of ion chromatographic systems. In mass spectrometry, ions are produced from a sample that will be examined. After that, these ions are separated and identified quantitatively. On the basis of distinct trajectories of moving ions with different mass/charge (m/z) ratios in magnetic or electrical fields, separation is accomplished.

Early 20th-century experiments and research that sought to understand how charged particles behaved in magnetic and electrostatic force fields gave rise to mass-spectrometry. A. J. Dempster directional focusing (1918), J. J. Thompson investigation into the behavior of ionic beams in electrical and magnetic fields (1912), and F. W. Aston energy focusing are well-known names from these early days (1919).

In this manner, a technique improvement was made that made it possible to gather crucial data on the isotopes' natural abundance. The first reliable commercial mass spectrometers were then built in the early 1940s, which was followed by the initial analytical applications. This was primarily done to quantitatively determine the various components of complicated crude oil blends.

Beginning in the early 1960s, mass spectrometry was used to identify and unravel the structures of more complex organic substances, such as polymers and biomolecules. Since then, the method has evolved into a potent and adaptable instrument for this purpose, providing data that partially overlaps and complements that of other methods, such as NMR.

It may come as a surprise that a method that at first glance does not seem to provide any more information than a particle's weight could be so crucial given that it is hard to think of a more prosaic attribute of a molecule than its molecular weight. The controlled fragmentation of the early molecule ions produces intriguing data that can aid in the understanding of structure. Additionally, the weights may now be calculated with enough precision to yield elemental compositions.

5.2. PRINCIPLES OF MASS SPECTROMETRY

Multiple ions are produced from the sample being studied by a mass spectrometer, which then separates them based on their unique mass-to-charge ratios (m/z) and logs the relative abundance of each ion type. The creation of gas phase ions of the molecule, primarily through electron ionization, is the initial step in the mass spectrometric study of substances.

This molecular ion splits into pieces. Fragmentation occurs in turn for each major product ion formed from the molecular ion, and so forth. The mass spectrometer separates the ions based on their mass-to-charge ratio, and they are detected proportionally to their abundance. This results in the creation of the molecule's mass spectrum.

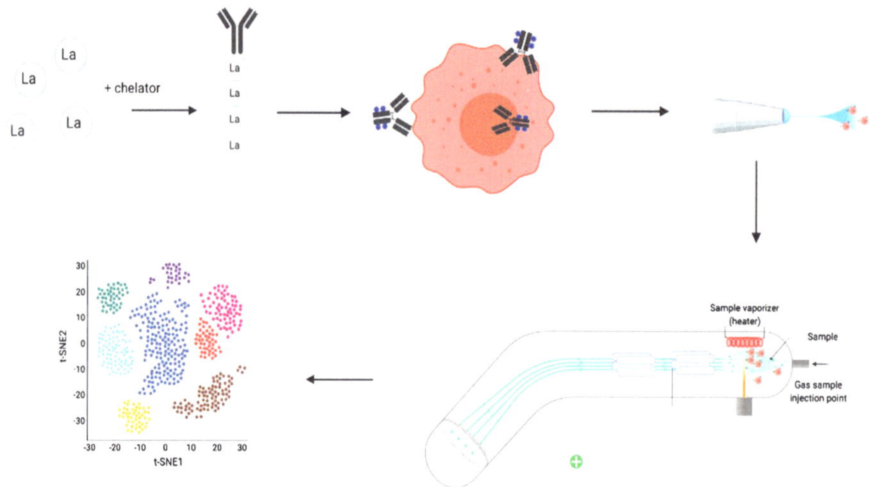

Figure 5.1. Mass spectrometry principle.

Source: Image by Wikipedia Commons

The outcome is shown as a plot of ion abundance against the mass-to-charge ratio. Ions reveal details about the makeup and structure of their precursor molecule. The molecular ion, if present, reveals the molecular mass of a pure substance by appearing in the spectrum at the highest value of m/z (followed by ions with heavier isotopes).

Though the workings of a modern analytical mass-spectrometer are straightforward to comprehend, the same cannot be said for the actual device. One of the most complicated electronic and mechanical instruments a chemist may come across is a mass spectrometer, particularly a multi-sector instrument.

Consequently, this entails substantial expenses for both purchase and maintenance in addition to specific operator training (s).

5.2.1. Measurement Principles

The components that make up an analytical mass spectrometer are as follows, as is the process:

- Only a very small amount of a compound is evaporated, usually one micromole or less. A pressure of roughly 10-7 mbar is maintained in the ionization chamber, where the vapor is leaking.
- An electron beam is now used to ionize the molecules of the vapor. This beam is generated by a filament-heated cathode. Inductive effects rather than rigorous collision are used to produce ionization. Positive ions are formed mostly as a result of the loss of valence electrons.
- A small positive charge (a few Volts) applied to the repeller opposite the exit-slit forces the positive ions out of the ionization chamber (A). The electric field (A>B) of several hundred to thousands of volts accelerates the ions as they exit the ionization chamber before they enter the analyzer.
- At a pressure of roughly 10-8 mbar, ion separation occurs in the analyzer, in this case, a magnetic sector. The ions are moved in a direction perpendicular to a high magnetic field. Due to the Lorentz acceleration, the fast-moving ions will then travel in a circle whose radius depends on the ion's mass/charge ratio and the strength of the magnetic field. By altering the accelerating voltage (A>B) or the magnetic-field force, ions with various mass/charge ratios are driven through the exit slit.

- The ions collide on a collector electrode after passing the exit slit. The resulting current is amplified and registered depending on whether an accelerating voltage or magnetic field is present.

The fact that an excess of energy forms a variety of positive ions after an electron interacts with a particular molecule makes mass spectrometry applicable to the identification of substances. The resulting mass distribution serves as the identifying feature (fingerprint) of the particular molecule.

There are some similarities between this and IR and NMR. Because information is given in terms of the masses of structure-components, mass-spectrograms are in some ways simpler to read.

The molecular weight of the particles can be ascertained using mass spectrometry (MS), an analytical technique that differentiates ionized particles like atoms, molecules, and clusters by exploiting variations in the ratios of their charges to their respective masses (mass/charge; m/z).

The components of an MS instrument are as follows: an ion source, which separates the sample molecules into ions; a mass analyzer, which sorts the ions according to their masses by using electromagnetic fields; a detector, that measures the value of an indicator quantity and thus gives information for determining the abundances of each ion present; and a computer, which controls the mass analyzer and manages the data obtained from the detector.

The ion source and mass analyzer techniques that are commonly used in these modules come in a variety of forms, and choosing the right one for IMS is crucial.

MALDI is one of the most often utilized IMS techniques out of all the other ionization techniques. With the help of a matrix and ionizing light, a large variety of biomolecules, ranging in size from tiny metabolite molecules (m/z 1000) to much larger proteins with molecular weights of 105 (Chaurand et al. 2006; Stoeckli et al. 2002), can be analyzed.

The principle behind MALDI was created by Koichi Tanaka, who won the Nobel Prize in Chemistry in 2002. It was further developed as proteins and DNA are examples of high-molecular-weight compounds that are brittle and prone to fragmentation when ionized using traditional ionization techniques.

The biomolecules are shielded from the direct laser beam and made more vaporizing and ionizing by a matrix made of crystalline molecules such as 2,5-dihydroxybenzoic acid (DHB) and sinapinic acid (SA). For successful mass spectra, the matrix needs to be selected properly.

The time-of-flight (TOF) analyzer for IMS is the most versatile. The vacuum analyzer separates the pulsatile ions produced by the ion source according to their time of flight, with lighter ions reaching the detectors faster than heavier ones.

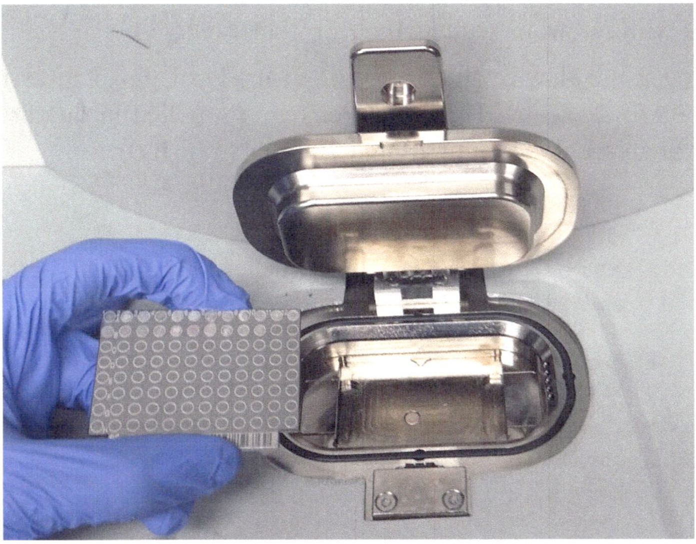

Figure 5.2. MALDI-TOF target plate for microbial identification.

Source: Image by Wikimedia Commons

As a result, the machine can calculate the atomic or molecular weight of each ion. A turbomolecular pump and a diffusion pump typically maintain a high vacuum within a mass analyzer since additional gas molecules could scatter the sample ions during the flight. By using tandem mass spectrometry, often known as MSn, it is possible to examine a molecule's structural makeup.

For instance, in MS/MS, isolated ionized molecules produced by MS1 are deteriorated and fractured through collision with a noble gas or through other means, and the pieces are then examined by MS2.

By comparing the molecular weights of the compound and the fragmented ions acquired by MSn with those in an existing database, a compound's molecular identity can be determined. A big molecule like a protein can be broken down into its component peptides for analysis.

5.2.2. The Ion Source

Ionization chamber, ion source, and often electron impact scheme. Ions of the substance under investigation can be created in the chamber in a variety of methods. The most typical method involves irradiating sample vapor molecules with electrons with a 70 eV or higher energy.

Heating a metal wire (filament), often made of tungsten or rhenium, produces these electrons. These electrons are propelled in the direction of the anode at an acceleration of around 70 Volts (5 to 100).

Figure 5.3. Ionization chamber made by Pierre Curie, c 1895-1900.

Source: Image by Wikimedia Commons

One or more electrons from the neutral molecule may be lost during the bombardment, forming positively charged molecular radical-ions. Just one out of every 103 molecules in the source is ionized. Although different compounds have different ionization probabilities, it has been discovered that the cross-section for the majority of molecules is at its highest for electron energies between around 50 and 100 eV.

Since sensitivity is near to a maximum and fragmentation is unaffected by little variations in electron energy around this value, the majority of current collections of electron impact spectra are based on spectra collected with electrons of roughly 70 eV. The radical ions typically gain excess energy during this ionization, breaking one or more bonds and creating fragment ions.

A molecule's potential for fragmentation: It should be noted that this is a simplified explanation and that there are many different ways to create fragments in real life, including through re-arrangement reactions. It is also conceivable for a molecular radical cation to split into a neutral molecule and a fresh fragment radical cation.

5.3. USES OF MASS SPECTROMETRY IN BIOCHEMISTRY

Biochemically, variations in protein structure are frequently quite significant. Many methods used to examine proteins may not be able to appropriately separate these variants from one another. However, it is possible to distinguish proteins that may only differ from one another by a few AMU because of the high accuracy of mass determination provided by MS analysis.

Additionally, because mass values have a direct correlation with particular chemical structures, MS analysis enables interpretations that could be challenging using other analytical techniques.

The finding that some proteins' cDNA-derived sequences were off was one of the earliest examples of the value of MS protein analysis. The additional examples that demonstrate the flexibility of MS in structural studies of biomacromolecules are discussed in the sections that follow.

5.3.1. MS and Microheterogeneity in Proteins

Many proteins are improperly, insufficiently, or otherwise erroneously digested when expressed in heterologous expression systems (such as a human gene cloned into and expressed in *Escherichia coli*).

This results from the possibility that the expression system is deficient in certain enzymes needed for processing catalysis or from the possibility that these enzymes are simply unable to detect processing sites as a result of species-specific sequence variations. This could take place anywhere along the polypeptide chain, including its termini.

Eukaryotic proteins are biosynthesized with the starting methionine intact; this methionine is later digested by enzymes to eliminate it post-translationally. In contrast, the biosynthesis of prokaryotic proteins begins with an N-formyl methionine.

It's possible that the bacterial and eukaryotic systems, respectively, lack the proper enzymes for removing initiation residues from prokaryotic and eukaryotic proteins. The N-terminus is also capable of a variety of different chemical configurations, which may cause N-terminal "blocking" (i.e., an inability to sequence the protein chemically using the Edman degradation).

This may be significant, for instance, if regulatory authorities demand the N-terminal sequence as proof of a protein's identity or purity during the commercial manufacture of recombinant proteins. The polypeptide chain's C-terminus has the potential for heterogeneity as well.

In heterologous expression systems, protein production may be prematurely stopped, leading to a variety of C-terminal sequences known as "ragged ends." Furthermore, several exopeptidases may be present in bacterial cells and trigger proteolysis at the N-terminus or C-terminus of proteins/peptides during the purification process.

Mass variations brought on by covalent modifications are detectable by MS. We can therefore quickly recognize such alterations by comparing the m/z values of protein variations. The plasticity of MS is largely due to this determination's independence from the chemical underpinnings of heterogeneity.

5.3.2. Confirmation and Analysis of Peptide Synthesis

The creation of model peptides that can be employed in a range of research has proven to be made possible by solid-phase peptide synthesis (e.g., synthesis of epitopes for antibody preparation, design of novel antibiotics and identification of novel protein ligands in drug screening programs).

Utilizing amino acids that are first chemically protected or inhibited at -NH2 and other functional groups before being chemically activated at their COOH group is the technique. These amino acids will easily interact with an immobilized amino acid's exposed -NH2 group to establish a peptide bond.

A polypeptide chain of 20–30 residues can be created from the C-terminus by unblocking the blocked -NH2 group, allowing a second cycle of reaction to result in the production of a second peptide bond. After the peptide is fully synthesized, the block is removed, revealing a C-terminal -COOH, and the functional groups are unblocked.

Peptides can be routinely created by adding commercially available activated/blocked amino acid derivatives in the desired order. This procedure is easily automated. The creation of deletion peptides and insufficient deblocking are frequent issues in peptide synthesis that result in heterogeneity in the finished peptide product.

It is also feasible to modify certain residues specifically, such as methionine through air oxidation or serine through trifluoro acetylation. Due to inadequate peptide bond formation during a synthetic cycle, deletion peptides are produced, with a portion of the peptide lacking the residue necessary for that cycle.

Since deletion peptides will be deficient in a certain amino acid, MS analysis can quickly and accurately identify them. Functional amino acid side-chains that undergo incomplete deblocking may still retain their blocking groups, giving them an unsuitable covalent structure.

Since different amino acid side chains contain different weights of blocking groups, MS analysis can identify specific partial deblocking stages and help enhance them in subsequent rounds of synthesis, either by extending reaction periods or changing the environment under which reactions occur.

5.3.3. Peptide Mapping

A distinct population of peptides is produced when certain end proteinases digest proteins. High-resolution separation of these may be accomplished using reversed-phase HPLC with C-18 stationary phase.

This is especially helpful when comparing related proteins, such as isoenzymes or hemoglobin, because common peptides are predicted in cases where the fundamental structure of the proteins is somewhat similar. In addition to HPLC analysis per se, MS provides a method for peptide map analysis that provides molecular mass information about the peptides.

By mass comparison alone, it may be possible to link a peptide to a specific region of the protein's fundamental structure, albeit it's important to take into account the isobaric nature of some residues in such tests.

5.4. MASS SPECTROMETRY OF PROTEINS/PEPTIDES

5.4.1. Sample Preparation

The structural analysis of highly pure proteins and peptides as well as combinations of proteins and peptides has become particularly popular using MS.

It is possible to acquire detailed structural information on factors of protein structure such as amino acid sequence, chemical and post-translational modification (e.g., N-terminal modification, glycosylation), disulfide bridge formation, and to compare closely-related proteins (e.g., mutants/wild-type) by peptide mapping thanks to the extremely high mass accuracy of these techniques (0.01 percent).

In these researches, it is frequently required to pretreat samples with particular chemical agents before performing an MS analysis. Prior to MS analysis, the sample should be carefully prepared, using standard measures such as cleavage, carboxymethylation, or reduction, although care should be given to ensure that any chemicals employed in these processes do not interfere with the analysis.

The analytical capability of MS is substantially increased by its interface with other high-resolution methods, and sample entry to MS via HPLC or electrophoresis frequently makes it easier to remove contaminants such as buffer ingredients or chemical agents employed in sample preparation. In some circumstances, it can be preferable to perform a one-dimensional analysis on the sample right away using MS.

To prevent the formation of adduct ions with alkali metals, it is required to remove any salts from the sample before analyzing proteins or peptides by MS. The quantity of protonated quasimolecular ions would decrease as a result. This is easily accomplished using the sample's reversed-phase HPLC, gel filtering, or dialysis. Early peptide studies using EI/MS and CI/MS required significant derivatization by acylation or esterification, which made sample molecules volatile. However, this is no longer required when employing matrix-based ionization modes.

It is only necessary to combine the protein (dissolved in water) with a solution of the matrix for ionization modes employing liquid matrices. Before obtaining the spectrum, the combination of matrix and sample is allowed to dry on the probe for modes using solid matrices (such as MALDI).

5.4.2. MS Modes Used In the Study of Proteins/Peptides

FAB, PDI, MALDI, ESI, and other methods make it simple to obtain information on the entire protein's molecular mass. Using a TOF analyzer often makes it easier to determine very big masses. Fragmentation information can be collected by CID, as well as by chemically or enzymatically pretreating sample proteins, such as trypsin digestion followed by peptide mass analysis.

It could be possible to assign masses to each peptide produced in the digest if the protein's sequence is known. By integrating MS analysis with other high-resolution methods like HPLC and electrophoresis, another degree of resolution is introduced.

5.4.3. Fragmentation of Proteins/Peptides in MS Systems

The main difference between proteins and peptides is the type of side-chain, or R, which is an oligomer of repeating (-NH-CHR-CO-) units. Peptide bonds (-NH-CO-), which have lower bond energies than conventional (-N-C-) bonds, hold the amino acid residues together.

Peptide bonds are frequently disrupted, releasing generally intact amino acid residues, when proteins or peptides fragment (for example, as a result of heat or acid hydrolysis). However, in MS investigations (like FAB/MS), there may be fragmentation at sites other than peptide bonds, leading to a complicated pattern of ions.

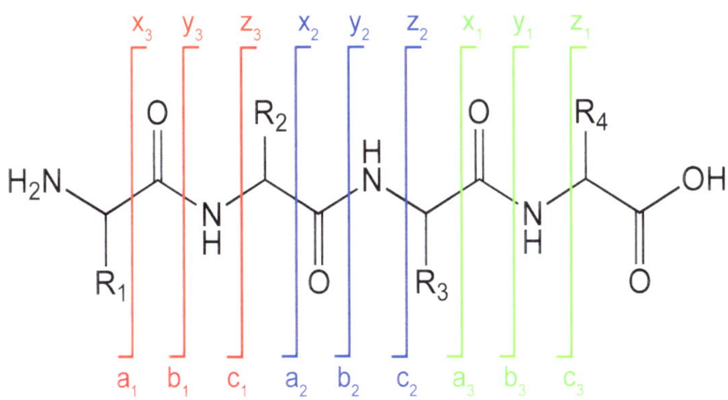

Figure 5.4. Peptide fragmentation.

Source: Image by Wikimedia Commons

Only 20 R-groups with known structures are frequently found in proteins, and because breakage points have set structural locations in relation to one another, they are related to one another by incremental m/z changes.

As a result, complex MS spectra can be interpreted to reveal information about the structure, particularly the amino acid sequence, of peptides and proteins. Roepstorff's terminology is frequently used to refer to polypeptide fragmentation in MS.

A polypeptide's MS-generated ions may have a positive or negative charge on either their C- or N-terminus. The fragment carrying the charge in that specific ion is indicated at the breakage site by a horizontal line pointing either towards the C- or N-terminus.

The sequence location of the residue is indicated by letters with a subscript number, n, which designates two series of potential fragments. The C-terminal ions are labeled as vn, wn, xn, yn, and zn (numbered from the C-terminus), while the N-terminal ions are designated as an, bn, cn, and dn (numbered from the N-terminus).

The sequence's first residue (minus the CO group) is represented by the ion designated as a1, the first two residues (minus the CO group) are represented by a2, and so on. The pairs a/x, b/y, and c/z signify main-chain breaking points that are the same (each member of a pair referring to a positive charge retained either on the N- or C-terminus, respectively).

The dn, vn, and wn breakage points are caused by side-chain fragmentations. By contacting samples with electrons via electron capture dissociation (ECD) and electron transfer dissociation (ETD), an alternate fragmentation technique is provided.

Both techniques rely on the magnetic field of an FTMS instrument and the interaction of electrons with peptide (or whole protein) cations. While radical anions are the source of electrons in ETD, the protonated sample anion interacts directly with a low-energy electron in ECD.

The generation of peptide/protein cations with an extra electron leads to rearrangement and backbone fragmentation. Importantly, side-chains do not have any bond fragmentation. As a result, unlike collision-based fragmentation, ECD/ETD are less sensitive to post-translational modifications and peptide length (collision-based fragmentation is poor with both big and small tryptic peptides) (these are often labile in collision-based dissociations).

5.5. INTRODUCTION TO SPECTROSCOPIC TECHNIQUES

To investigate the molecule-level interaction, spectroscopic methods are used. In this regard, nuclear magnetic resonance (NMR) is one method. NMR is utilized to analyze the interaction between mucus and polymers, and carbon and/or proton NMR can also elaborate on this interaction.

For this method, there is no need for preparation or labeling. It is based on the idea that the interaction of polymers with mucus is highly sensitive to the mucus' great affinity for tiny, high-affinity probes like ketotifen fumarate.

Therefore, the ability of mucin to bind to a probe changes depending on whether or not mucoadhesive polymers are present. Uccello-Barretta et al. used this technique to study the mucoadhesive properties of tamarind seed polysaccharide, arabinogalactan, and a mixture of tamarind seed polysaccharide and hyaluronic acid with bovine submaxillary mucin.

They discovered that selective proton relaxation rates are highly responsive to such interaction. This method, however, works well with nonionic polymers that exhibit noncovalent bonding.

Polymer chain conformations can be observed using a variety of spectroscopic methods. To determine the degree of helicity and investigate other facets of protein and nucleic acid tertiary structure, ultraviolet spectroscopy (UV) is almost solely utilized for biopolymers. For instance, when helical conformations are produced, hypochromic effects can be seen in the UV spectra of nucleic acids and nucleotides.

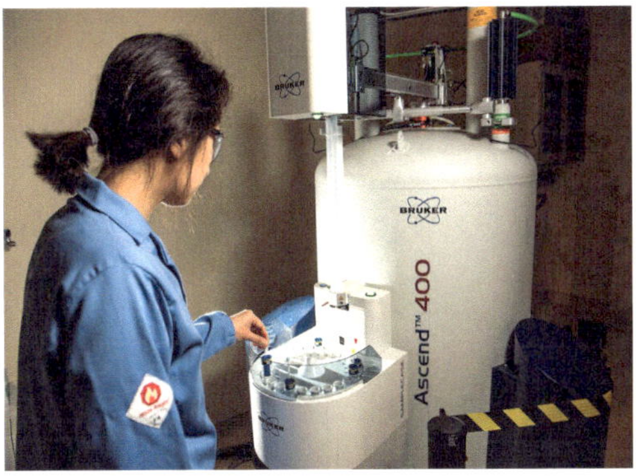

Figure 5.5. The new Nuclear Magnetic Resonance (NMR) instrument in BSF analyzes small molecular samples.

Source: Image by Flickr

There is some UV proof that isotactic polystyrene partially maintains a helical shape in solution in terms of synthetic polymers. Measurements with infrared technology support this.

The following section will go over how infrared spectroscopy (IR) is generally used to characterize polymer conformations in bulk samples. Since the skeletal vibration frequencies are conformationally sensitive, it can also be employed to research solution state conformations. The trademark bands for the tt, tg, and lengthier sequences can be found.

Researchers have discovered helical sequences in isotactic polystyrene using these conformationally sensitive IR bands, and they have also demonstrated the lack of cooperative helix-coil transitions in this substance. The study of hydrogen bonding interactions in biopolymers and synthetic polymers can both benefit greatly from IR spectroscopy.

Information from Raman spectroscopy is complementary to that from IR measurements. As we've already seen, a vibrational mode's dipole moment must change in order for it to be IR active. As a result, IR-active groups include single bond-OH, single bond-NH, and single bond-CH bonds. In fact, the single bond-OH infrared bands are so potent that they frequently make it problematic for some IR experiments to use water as a solvent.

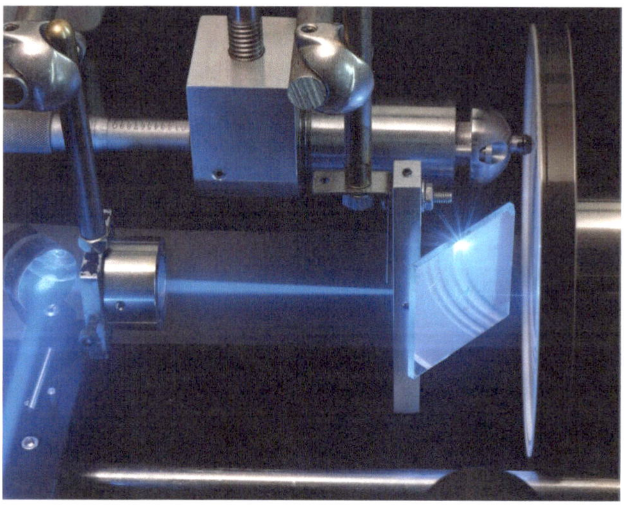

Figure 5.6. Raman spectroscopy enables scientists to study at the molecular level the chemical and physical changes of ceramic materials as they undergo friction and wear degradation.

Source: Image by Flickr

A symmetrical bond, like a carbon-carbon bond or a single bond, has a stretching vibration that is IR-inactive. However, a strong Raman band is produced by the C-single Bond Stretching vibration. Raman investigations can also be conducted in aqueous solvents, which makes Raman spectroscopy a perfect technique for examining the conformations of biological macromolecules.

With regard to the conformational changes that take place in heme-proteins upon binding tiny ligands, Raman spectroscopy offers a wealth of fresh knowledge.

The conformations of biopolymers have long been studied using fluorescence measurements. Tyrosine, phenylalanine, and tryptophan all glow. Solvent studies may, in some circumstances, be utilized to establish the hydrophobicity of the environment around these residues since solvents have an impact on the fluorescence intensity of proteins. To add more conformationally sensitive information, fluorescent dyes or fluorescence probes can be entrained in or connected to biological molecules.

Examining the conformations of synthetic polymers in solution makes excellent use of studies on excimer production. An association complex between an excited and a ground-state chromophore is known as an excimer. In contrast to what would often be seen, the excimer emission band is pushed toward longer wavelengths.

The spatial closeness of the chromophores is necessary for the efficient production of excimer complexes and can be utilized to determine conformation. Since excimer synthesis, in some circumstances, necessitates that a conformational transition takes place within the fluorescence lifetime, excimer fluorescence can also be utilized to investigate conformational mobility in polymer chains.

The measurement of electromagnetic energy intensity as a function of wavelength is known as optical spectroscopy, and it deals with electromagnetic radiation (light) in the ultraviolet, visible, and near-infrared ranges. We refer to this as optical spectroscopy since optical instruments are used to focus or disperse light in these procedures. It's an incredibly adaptable science with a variety of methods for gleaning knowledge from light interacting with the surroundings. Depending on the application and the type of data needed, the best spectroscopic approach will be chosen.

The type of interaction between energy and the material to be sampled is frequently used to categorize spectroscopy techniques. The oscillating electric and magnetic forces that form electromagnetic radiation radiate outward along a linear route at a constant speed. These electromagnetic waves can be described in terms of their wavelength, amplitude, frequency, intensity, polarization, and other properties.

The characteristics of the radiation alter when it interacts with different materials. We gather useful information by detecting radiation under various experimental conditions. Numerous spectroscopic methods examine how light and matter interact at the atomic level.

Working with "free atoms" may be required in this, as in laser-induced breakdown spectroscopy, when a laser ablates a sample surface to produce a micro-plume of plasma for elemental analysis. Additionally, light scattering resonances may be involved, such as a measurement of Raman scattering that provides insight into the atomic composition of a sample.

5.6. FLUORESCENCE SPECTROSCOPY

A quick and sensitive technique for describing molecular environments and events in samples is fluorescence spectroscopy. Fluorimetry is preferred over other analytical methods because of its exceptional sensitivity, high specificity, ease of use, and low cost.

It is a well-known and effective approach that is applied in a wide range of biotechnology, forensic, environmental, industrial, and medical diagnostic applications. Both quantitative and qualitative analyses can benefit from using this useful analytical tool.

Based on the fluorescent properties of an analyte, fluorescence spectrometry is a quick, easy, and affordable way to estimate the concentration of the analyte in solution.

When the type of compound to be examined (the "analyte") is known, it can be used for relatively straightforward analyses to perform a quantitative analysis to ascertain the concentration of the analytes. Fluorescence is mostly used to measure substances in solution.

A beam with a wavelength ranging from 180 to 800 nm passes through a solution in a cuvette during fluorescence spectroscopy. The light that the sample emits is then measured at an angle. Fluorescence spectrometry allows for the measurement of both the excitation spectrum—the light that is absorbed by the sample—and the emission spectrum—the light that the sample emits. The intensity of the emission is directly proportional to the analyte concentration.

Alzheimer's Photon emission processes called fluorescence and phosphorescence take place when molecules relax from electronic excited states. Transitions between the electronic and vibrational states of polyatomic fluorescent molecules take place during these photonic processes (fluorophores).

The main component of fluorescence spectroscopy is fluorophores. Tyrosine, Tryptophan, Fluorescein, and other molecules with aromatic rings are the most common fluorophores.

A substance emitting light without being heated is said to be luminescent, making it a type of cold body radiation. Chemical interactions, electrical energy, subatomic vibrations, or stress on a crystal are all potential causes. Two conditions must be met for luminescence:

- A semiconductor structure with a non-zero band gap is required for the luminous material. [Metals without a band gap do not produce luminescence.]
- Before luminescence can occur, this material needs to receive the energy.

5.5.1. Fluorescence Spectroscopy Configuration

Fluorescence spectroscopy involves passing a light beam with a particular wavelength band through a solution, which then emits the light in the direction of a filter and a detector for measurement. It is possible to measure how much light is absorbed by the sample (excitation spectrum) and how much light is released by the sample (emission spectrum).

Since the concentration of the analyte component in the solution is inversely proportional to the emission spectrum, the concentration levels can be identified.

5.5.2. Fluorescence Spectroscopy Filters

When describing the configuration of a fluorescence spectroscopy device, we briefly covered filters, and the majority of these devices are built around the filters they contain.

Here, we'll go through a few of the several filter types that can be applied to a fluorescence spectroscopy device.

With almost negligible absorption, an interference filter transmits one wavelength range while reflecting another. These filters serve as a wavelength selector in a fluorescence spectroscopy device by reflecting the unwanted wavelengths. Bandpass filters are one of the many variations of these filters.

When a bandpass filter is in operation, only specific wavelengths can pass through while all other wavelengths are blocked.

These fluorescence spectroscopy devices measure the excitation and emission spectra using bandpass filters. They are employed to eliminate any additional sources of excitation or emission that might influence the outcomes.

To divide incident light into different wavelengths or by intensity, a beamsplitter is utilized. A dichroic beamsplitter is employed in a fluorescence spectroscopy apparatus to reflect and transmit light into the bandpass filters.

5.7. FLUORESCENCE CORRELATION SPECTROSCOPY

The fundamental idea behind fluorescence correlation spectroscopy (FCS) is that a fluorescing molecule exhibits a particular free diffusion velocity that is closely associated with its size. Therefore, a larger molecule will diffuse through a given spherical volume more slowly.

In FCS, this fundamental property of molecules is used to investigate a variety of topics, including protein-protein interactions, attachment, and more. Fluorescence Correlation Spectroscopy (FCS) studies dynamic molecular events, such as diffusion or conformational variations of biological molecules or synthetic particles, by using statistical deviations of the fluctuations in fluorescence.

The auto correlation function (ACF) is primarily used to determine the quantity and rate of diffusion of fluorescent particles within the focus volume. All of these characteristics of FCS make it a superb diagnostic and research tool for a variety of serious illnesses. FCS is an excellent tool for understanding the many pathophysiological processes connected to microbial infectious illnesses because of its many unique features.

5.7.1. Diagnostic Applications of Fluorescence Spectroscopy

Fluorescence spectroscopy has been successfully used in numerous studies as a diagnostic tool for various bacteria at the genus, species, and group level through the use of spectral fingerprints.

A collection of oral bacteria called black pigmented bacilli underwent spectral investigations that revealed a considerable variation in the spectral signature of each bacterium. The fluorescent profiles of the bacteria that cause pediatric otitis media viz. *S. pneumoniae*, *S. aureus*, *M. catarrhalis*, and *H. influenzae* have been investigated. These investigations demonstrated that every bacterium produces a distinctive fluorescence characteristic.

According to the statistics, it might be a very good non-invasive fluorescence-based diagnostic method for otitis media. Another study used autofluorescence spectrum differences in conjunction with the Principal Components Analysis (PCA) technique to quickly identify three different bacterial species (*Escherichia coli*, *Enterococcus faecalis*, and *Staphylococcus aureus*).

These investigations suggested that bacteria can be quickly identified with greater than 90% sensitivity and specificity.

5.7.2. Measurement of Fluorescence and Chemiluminescence

A spectrofluorimeter is used to measure fluorescence. A sample cuvette containing the fluor is passed through by an incident beam of radiation of a specific wavelength. A photomultiplier tube detects the radiation that is being emitted, which has a longer wavelength than the incident beam. The spectrophotometer's design is often similar to this.

The two primary distinctions are that a second monochromator is needed to select for the different wavelengths of the emitted light and that released radiation is detected at 90 degrees to the direction of the incident light beam. This design prevents accidental detection of the incident beam because fluorescence is released from the fluor in all directions.

This design has the effect of requiring cuvettes used for fluorescence measurements to have four transparent sides, but cuvettes used for absorbance measurements frequently have frosted or blackened side-walls.

A luminometer is used to measure chemiluminescence. The light-emitting process takes place in a cuvette, where the wavelength and power of the light are measured. The produced light can be detected on photographic film when used as a component of an in-situ detection method (for example, western blotting).

Since some energy may be lost during vibrational transitions, only a part of the light energy that was initially absorbed is actually released as radiation. Two additional procedures can reduce or quench the light energy that is emitted from the material. Internal quenching results from a structural characteristic of the stimulated molecule, such as a structural rearrangement.

Either the excited molecule interacts with another molecule in the sample or a different chromophore in the sample absorbs the excited or released light, which causes external quenching. Nonradiative energy loss results from all types of quenching.

Since both internal quenching and vibrational transitions originate from chemical structure, the amount of energy lost owing to each will depend on the specific molecule. Impurities in the preparations may be the cause of external quenching, or conversely, the experimenter may have purposefully added the contaminants. Examples of compounds that can serve as external quenchers include acrylamide, iodide, and ascorbic acid.

5.8. INFRARED SPECTROSCOPY

One of the most popular spectroscopic methods used by organic and inorganic chemists is infrared (IR) spectroscopy. It is simply the measurement of distinct IR frequency absorption by a sample that is placed in the path of an IR beam. Finding the sample's chemical functional groups is the major objective of IR spectroscopic analysis.

Different functional groups absorb IR radiation at particular frequencies. IR spectrometers can accept a variety of sample types, including gases, liquids, and solids, using various sampling accessories. Sir William Herschel made the discovery of infrared radiation in 1800.

Herschel was looking into the energies connected to the visible spectrum of light's wavelengths. The well-known visible spectrum of the rainbow hues, or the visible spectrum from blue to red with the equivalent wavelengths or frequencies, was visible when sunlight was focused through a prism.

The investigation of electromagnetic (EM) wave interactions with matter is known as spectroscopy. The energy levels of the rainbow hues are matched to the color wavelengths. Herschel noted that the temperature rose from the blue to the red region of the visible spectrum by slowly moving the thermometer through the spectrum and monitoring the temperatures along the way. Herschel then made the decision to take a temperature reading just below the red region, believing that the increase in temperature would stop outside of the visible spectrum. To his surprise, however, he discovered that the temperature was actually higher than he had anticipated.

These rays, which were below the red rays, were referred to by him as "non colorific rays" or invisible rays. Later, these rays came to be known as "infrared rays" or IR light. The human eye cannot see this light.

Wavelengths between 390 and 750 nm will elicit a response from a typical human eye. At 0.75 nm, the IR spectrum is initiated. A nanometer (nm) is 10-9 meters. Near infrared (NIRS), mid infrared (MIRS), and far infrared are the three sub bands of the infrared spectrum (FIRS).

5.8.1. Principle of Infrared Spectroscopy

The foundation of IR spectroscopy is the idea that molecules absorb particular frequencies that are unique to their structure. All of the atoms in molecules are continuously vibrating with respect to one another at temperatures higher than absolute zero. A beam of IR light is sent across a sample to record its IR spectra.

A molecule absorbs IR radiation when the frequency of a particular vibration is equal to the frequency of the IR radiation being directed at the molecule. The amount of energy absorbed at each frequency can be determined by looking at the transmitted light (or wavelength). IR spectrometers can accept a variety of sample types, including gases, liquids, and solids, using various sampling accessories.

5.8.2. Equipment of Infrared Spectroscopy

Three essential parts make up an IR spectrometer: a radiation source, a monochromator, and a detector. An inert substance heated electrically to 1000°C–1800°C, such as rare-earth oxides, silicon carbide, or nichrome coil, serves as the standard radiation source for IR spectrometers.

The monochromator is a device that disperses a wide spectrum of light and provides an ongoing calibration series of electromagnetic energy bands with determinable wavelengths or frequency ranges. The dispersive elements employed in conjunction with variable-slit mechanisms, mirrors, and filters are prisms or gratings.

For instance, a rotating grating can concentrate a specific frequency range on a mechanical slit. Better resolution is made possible by narrower slits, while increased system sensitivity is achieved by wider slits. Thermocouples, thermistors, and Golay detectors are examples of detectors that gauge the heating impact brought on by infrared radiation.

Expansion of a non-absorbing gas (Golay detector), electrical resistance (thermistor), and voltage at the junction of different metals (thermocouple) are only a few examples of the physical property changes that can be quantitatively measured.

5.8.3. Types of IR Spectrometers

Dispersive IR (DIR) spectrometers and Fourier transform IR (FTIR) spectrometers are the two main types of spectrometers used in IR spectroscopy.

A monochromator divides the radiation from a broadband source that passes through the sample in a standard dispersive IR spectrometer into its component frequencies. The beams then strike the detector, which produces an electrical signal and a response from the recorder. The majority of dispersive spectrometers are double-beam instruments.

The sample and reference chambers are each traversed by two identical beams from the same source. The reference and sample beams are alternately focused on the detector using an optical chopper (like a sector mirror).

Usually, the change in IR radiation intensity caused by the sample's absorption is identified as an off-null signal, which synchronous motors translate into the recorder response.

Due to their improved speed and sensitivity, Fourier transform infrared (FTIR) spectrometers have lately supplanted dispersive instruments in the majority of applications. In FTIR spectroscopy, all component frequencies are viewed concurrently as opposed to sequentially, as in a dispersive IR spectrometer.

Figure 5.7. The Fourier Transform Infrared Spectroscopy (FTIR) Chamber.

Source: Image by Flickr

Radiation source, interferometer, and detector make up the fundamental three parts of a spectrometer in an FT system. Both dispersive and Fourier transform spectrometers use the same kinds of radiation sources. To improve power and stability, the source is more frequently water-cooled in FTIR instruments. Interference signals created by the interferometer contain infrared spectral data that was generated after passing through a sample.

The moving mirror, stationary mirror, and beam splitter are the three active parts of the interferometer. The interferogram output is created when the mirror is moved at a constant speed while the intensity of radiation reaching the detector varies in a sinusoidal fashion. The interferogram is transformed into the final IR spectrum via a mathematical procedure called the Fourier transform.

5.9. SPECTROSCOPIC TECHNIQUES USING PLANE-POLARIZED LIGHT

5.9.1. Polarized Light

The waves that make up plane polarized light all have the same direction of vibration. As the wave moves forward with circular polarization, the electric vector rotates about the direction of propagation. Reflection or passing through filters, such as certain crystals, that transmit vibration in one plane but not in others, can polarize light.

Electric and magnetic vectors that are perpendicular to one another make up electromagnetic radiation. The electric vectors in a beam of radiation coming from the sun or any other light source are distributed at random around the beam axis.

However, a beam of plane polarized or linearly polarized light would be produced if the electric vectors in one plane could be chosen. This can be accomplished by running unpolarized light through a polaroid filter. One way to conceptualize plane polarized light is as two equally strong but oppositely polarized circular vectors as left and right circularly polarized components. A helical path would be described by the electric vector if we followed it down the axis of a circularly polarized light beam while maintaining its equal magnitude.

5.9.2. Chirality in Biomolecules

When bound to four different chemical groups, the four valencies of carbon can take on two alternative isomeric forms, d and l. Enantiomers, also known as optical isomers, are pairs of isomers that cannot be superimposed as mirror images of one another.

The C atom of the amino acid alanine is referred to as an optical center or a center of asymmetry to indicate that the two isomers develop from various configurations around this atom. There could be four, six, or more potential stereoisomers in some compounds when there is more than one optical center.

Although they are not connected to one another as enantiomers, several of these structural variations are clearly stereoisomers (i.e., nonsuperimposable mirror images of each other). Diastereoisomers are what these are. Enantiomeric pairings' existence is referred to as chirality, and molecules that can do so are referred to as chiral molecules.

Because hands are excellent examples of items that are not superimposable mirror images, the word "chiral" is derived from the Greek word for hand (cheir). Conventionally, the l- and d-enantiomers are designated by comparison with the reference chemical l-glyceraldehyde. An alternative notation known as the RS convention gives the absolute configuration of enantiomers. The fact that many biomolecules are chiral makes the phenomena of chirality particularly significant in the field of biochemistry. In biological systems, one of a pair of enantiomers is typically chosen.

For instance, the majority of amino acids in living systems are l-en4antiomers, while d-enantiomers only very infrequently exist (e.g., in antibiotic peptides). The majority of monosaccharides are also d-enantiomers, with l-enantiomers only sometimes occurring. On the other hand, the majority of chemical reactions that take place in solutions produce a 50:50 combination of d- and l-enantiomers.

Chirality affects the structure, form, and functional characteristics of biomolecules in a number of significant ways. For instance, proteins' right-handed helices are produced by l-amino acids. D-amino acids produce left-handed helices that are mirror images of right-handed helices, but heteropolymers made of a combination of d- and l-amino acids are incapable of producing any helices.

Proteases can hydrolyze peptide bonds in polypeptides made entirely of l-amino acid substrates but not in those made entirely of d-amino acid substrates. Conversely, all-d polypeptide substrates can be broken down by synthetic proteases generated chemically from all-d amino acids (such as the HIV protease), but not those made from all-l amino acids.

In fact, the artificial HIV protease mirrors the natural all-l enzyme, demonstrating that chirality is prcscrvcd throughout the protein structural hierarchy. A pair of enantiomers has also been found to have different levels of toxicity toward humans, with one enantiomer being hazardous while the other is harmless.

Thalidomide, a fertility medicine that was extensively administered in the late 1950s, is a well-known example of this. One of these compound's enantiomers was benign, while the other was teratogenic, causing deformities or limblessness in offspring born to people who took the medication.

Because of this, the pharmaceutical industry is very interested in the potential application of enzymes for the enantioselective manufacture of novel medications.

5.9.3. Linear Dichroism (LD)

Individual molecules undergo the electronic transitions necessary for UV and visible light absorption along a certain axis. To ensure that the sample will produce the same absorption spectra regardless of the direction of an incident light beam, molecules that have been dissolved in an aqueous solvent are randomly distributed.

However, depending on the direction of the light beam, different absorption spectra would be observed if the molecules were all oriented in the same way. This characteristic is taken advantage of in linear dichroism by exposing oriented samples to plane polarized light. There are shown extreme situations where the incident beam is perpendicular to the transition axis and parallel to the axis of a certain transition.

In LD spectroscopy, samples are oriented along a single axis with respect to the incident linearly polarized light, and the relative orientation of the sample and light can be changed so that the incident light is either parallel to the axis of the sample or perpendicular to it.

By absorbing a sample onto a polymer (such polyvinyl alcohol) either before or after stretching it in a specific direction, the stretched polymer approach allows for orientation to be established. The sample molecule's long axis is aligned with the stretch direction. Squeezed gel orientation, electric field orientation, and flow orientation are further common methods.

5.10. CONCLUSION

This chapter discussed the basic significance of mass spectrometry. It also discussed the various principles of mass spectrometry. In this chapter, several applications of mass spectrometry in biochemistry have also been discussed.

Towards the end of the chapter, it discussed the mass spectrometry of proteins and peptides such as sample presentation, and MS modes used in the study of proteins and peptides. In this chapter, an introduction to spectroscopic techniques such as fluorescence spectroscopy, and infrared spectroscopy has also been discussed.

REFERENCES

1. B, T., n.d. *Infrared Spectroscopy: Principle and Types | Soil Mineralogy*. [online] Soil Management India. Available at: <https://www.soilmanagementindia.com/soil-mineralogy-2/infrared-spectroscopy-principle-and-types-soil-mineralogy/13268#:~:text=Principle%20of%20Infrared%20Spectroscopy%3A&text=When%20the%20frequency%20of%20a,each%20frequency%20(or%20wavelength).> [Accessed 4 July 2022].
2. Murayama, C., Kimura, Y. and Setou, M., 2009. Imaging mass spectrometry: principle and application. *Biophysical Reviews*, [online] 1(3), pp.131-139. Available at: <https://www.ncbi.nlm.nih.gov/pmc/articles/PMC5425681/#:~:text=Principle%20of%20MS,molecular%20weight%20of%20the%20particles.> [Accessed 4 July 2022].
3. n.d. *Basic Mass Spectrometry*. [ebook] www.whitman.edu. Available at: <https://www.whitman.edu/chemistry/edusolns_software/GC_LC_CE_MS_2017/CH%205a%202017.pdf> [Accessed 4 July 2022].
4. Poole, C., Prozorov, R., Farach, H. and Creswick, R., 2014. Spectroscopic properties. *Superconductivity*, [online] pp.577-650. Available at: <https://www.sciencedirect.com/science/article/pii/B9780124095090000147> [Accessed 4 July 2022].
5. Shahzad, A., Köhler, G., Knapp, M., Gaubitzer, E., Puchinger, M. and Edetsberger, M., 2009. Emerging applications of fluorescence spectroscopy in medical microbiology field. *Journal of Translational Medicine*, [online] 7(1). Available at: <https://translational-medicine.biomedcentral.com/articles/10.1186/1479-5876-7-99> [Accessed 4 July 2022].
6. Theophanides, T., 2012. Introduction to Infrared Spectroscopy. *Infrared Spectroscopy - Materials Science, Engineering and Technology*, [online] Available at: <https://www.intechopen.com/chapters/36273> [Accessed 4 July 2022].

CHAPTER 6

PRINCIPLES OF CLINICAL BIOCHEMISTRY

CONTENTS

Blood keeps a history of all the ways it has interacted with physiological and pathological processes. Homeostasis, the steady equilibrium that characterizes health, is made possible by the equilibrium of chemical reactions. Blood tests can detect abnormalities that endanger homeostasis, including specific electrolyte imbalances and the pH (acidity) of the blood that develops as a result of the disease.

6.1. INTRODUCTION

In order to diagnose and track the progression of a disease, clinical biochemistry refers to the study of the blood plasma (or serum) for a wide range of chemicals, including substrates, enzymes, hormones, and others. There is also an analysis of other bodily fluids such as urine, ascitic fluids, and CSF.

The following are basic checklists of variables impacting the most often requested analytes because one test is very rarely relevant to just one clinical condition. Therefore, a carefully chosen set of six tests can yield information pointing to a broad variety of various conditions by a process of pattern recognition, as opposed to six tests that only confirm or reject six possibilities.

Full hematological evaluations should be performed alongside biochemistry testing since doing so is necessary for the best identification of many of the most distinctive illness patterns. A list of potential diagnoses should be made before samples are taken based on the history and physical examination. The basic panel below can then be supplemented with other applicable tests.

Creating a differential diagnosis list based on the history and clinical examination is a necessary step in making a diagnosis. Tests can be chosen to include or omit as many of the differentials as possible based on this list.

The number of tests may increase until just one from the initial list is left to make the diagnosis. The list needs to be reevaluated if all differentials are eliminated. Unless the animal appears without clear-cut clinical signs, it is not a good idea to order testing without a reasonable differential list.

6.2. PRINCIPLES OF CLINICAL BIOCHEMICAL ANALYSIS

6.2.1. Basis of Analysis of Body Fluids for Diagnostic, Prognostic, and Monitoring Purposes

A change in the quantity or function of one or more proteins, which in turn causes changes in cellular, tissue, or organ function, lies at the root of the majority of human disorders. The biochemical makeup of body fluids typically changes significantly, which is a common indicator of malfunction.

A helpful tool for the identification and treatment of the prevailing illness state is the use of quantitative analytical biochemical tests on a wide variety of biological analytes in bodily fluids and tissues. We'll talk about the fundamental biological and analytical concepts that underlie these tests and how they relate to the fundamental concepts of quantitative chemical analysis.

Body fluids like blood, cerebrospinal fluid, and urine contain a lot of inorganic ions and organic molecules in both healthy and pathological states. While part of these chemical species' regular biological function is contained in that fluid, the bulk does not.

The presence of this latter category of chemical species in the fluid is caused by the release of cell components, particularly those found in the cytoplasm, into the surrounding extracellular fluid, which then enters the blood circulatory system.

This release is caused by both normal cellular secretory mechanisms and the temporal synthesis and turnover of specific cells and their organelles within the major organs of the body. This then transfers them to the liver, kidneys, and lungs, which are the primary excretory organs.

Ultimately, these cell components and/or their breakdown products are expelled in feces, urine, sweat, and expired air. Examples of cell components in this category include enzymes, hormones, intermediary metabolites, and small organic and inorganic ions.

Numerous factors that can be divided into three categories—chemical properties of the component, endogenous characteristics of the individual, and exogenous factors that are imposed on the individual—affect the concentration, amount, or activity of a specific cell component that can be detected in these fluids of a healthy person at any given time.

Chemical characteristics: In environments other than the normal cellular milieu, some molecules are inherently unstable. For instance, several enzymes depend on the presence of their cofactor and/or substrate for stability, and these components may be absent or present in insufficient quantities in the extracellular fluid.

It will also take little time for molecules to break down that can serve as substrates for catabolic enzymes prevalent in extracellular fluids, particularly blood. As a result, these two types of cell components have a short half-life outside the cell and are often present in low concentrations in fluids like blood.

Endogenous factors: Age, gender, body mass, and pregnancy are a few of these. For instance:

(a) Men have greater serum cholesterol levels than premenopausal women, but these differences disappear after menopause.

(b) Serum alkaline phosphatase activity is higher in children than in adults, and it increases in pregnant women.

(c) Obese people have greater serum insulin and triglyceride concentrations than thin people do.

(d) A metabolic product of creatine, serum creatinine, important in muscle metabolism is higher in individuals with a large muscle mass

(e) Between males and females serum sex hormone concentrations tend to differ and also change with age.

Exogenous factors: Time, activity level, caloric intake, and stress are examples of exogenous influences. Time-related hormone secretion occurs in a number of cases. As a result, the release of cortisol, thyroid-stimulating hormone (TSH), and to a lesser extent, prolactin exhibits a diurnal regularity.

When it comes to cortisol, its secretion peaks at about 9.00 am, drops throughout the day, and reaches a low between 11:00 pm and 5:00 am. Female sex hormone release changes throughout a menstrual cycle and 25-hydroxycholecalciferol (vitamin D3) secretion vary with the seasons, peaking in the late summer.

Immediately following a meal, the concentrations of glucose, triglycerides, and insulin increase in blood. A variety of hormones and neurotransmitters, including prolactin, cortisol, adrenocorticotropic hormone (ACTH), and adrenaline, can be secreted in response to stress, including the kind caused by the procedure of drawing blood sample by puncturing a vein (venipuncture).

Even in healthy individuals, there is a significant intra-individual variation (i.e., variation from one occasion to another) in the value of any chosen test analyte of diagnostic importance and an even greater inter-individual variation (i.e., variation between individuals) due to the influence of these various factors on the extent of release of cell components into extracellular fluids.

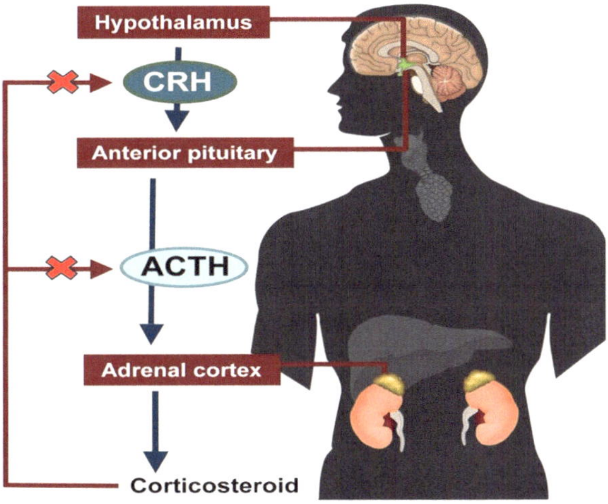

Figure 6.1. Hypothalamic-pituitary-adrenal axis. ACTH, adrenocorticotropic hormone; CRH, corticotropin-releasing hormone.

Source: Image by Wikimedia Commons

More significantly, the introduction of a disease state to these causes of intra- and inter-individual variation will produce much more variation between test occasions.

The integrity of cells in the affected organs is frequently compromised by many clinical disorders. This could cause the cells to become "leakier" or, in more extreme circumstances, to die (necrosis) and release their contents into the extracellular fluid surrounding them.

Most of the time, the degree of release of particular cell components into extracellular fluid compared to the healthy reference range will represent the degree of organ damage, and this relationship serves as the foundation for diagnostic clinical biochemistry.

As is the case in cirrhosis of the liver, for example, the mass of cells left to undergo necrosis will eventually diminish so that the release of cell components into the surrounding extracellular fluid will decrease even though organ cells are still being damaged.

If the cause of the organ damage persists for a long time and is essentially irreversible (i.e., the organ does not undergo self-repair). The measured amounts in this situation won't accurately depict how much organ damage has occurred.

A provisional clinical diagnosis based on the patient's medical history and physical examination has been created, and clinical biochemical tests have been devised to support it in four primary ways:

- To support or reject a provisional diagnosis by detecting and quantifying abnormal amounts of test analytes consistent with the diagnosis.

For instance, after a myocardial infarction (heart attack) that causes cell death in some heart tissue, serum levels of myoglobin, troponin-I (a component of the cardiac contractile muscle), creatine phosphokinase (particularly the CK-MB isoform), and aspartate transaminase all increase.

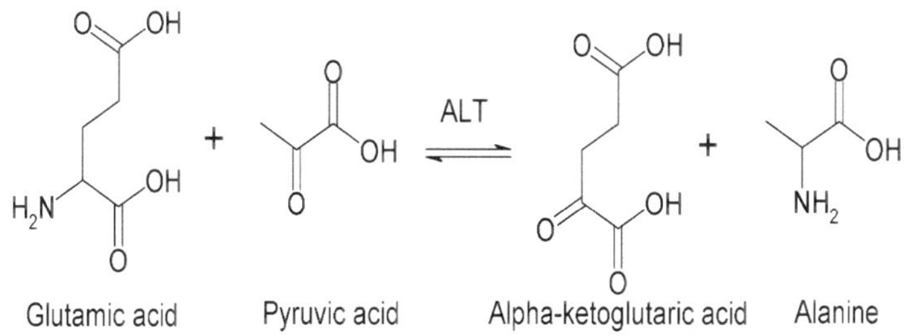

Figure 6.2. Alanine transaminase reaction.

Source: Image by Wikimedia Commons

The overall amount of released cellular components is amplified because the released cellular components also cause cell inflammation (leakiness) in neighboring cells. The results of tests can also be used to make a differential diagnosis, such as identifying the different types of jaundice (a condition in which the skin turns yellow due to the presence of bilirubin, a metabolite of hemoglobin), by measuring the activity of the enzymes alanine transaminase (ALT) and aspartate transaminase (AST), as well as by evaluating whether or not the bilirubin is conjugated with b-glucuronic acid.

- By doing the tests regularly and keeping an eye on the return of the test values to those within the reference range, one may keep track of recovery after therapy. For instance, the elevated blood enzyme activity mentioned above often recovers to reference range values within 10 days after myocardial infarction.

Similar to this, monitoring ovarian cancer treatment-related progress or recurrence can be done by measuring serum tumor markers like CA125.

- To check for elevated levels of important analytes to detect latent disease in people who appear to be healthy. For instance, testing immunoreactive trypsin for cystic fibrosis and serum glucose for diabetic mellitus.

In order to assess a person's risk of getting heart diseases, serum cholesterol levels are now frequently employed. This is crucial for people who have a history of the disease in their families.

The British Hyperlipidaemia Association has established an action limit of serum cholesterol >5.2 mM for individuals to receive clinical advice and counseling regarding the value of a healthy (low fat) diet and regular exercise and an even higher action limit of serum cholesterol >6.6 mM for the prescription of cholesterol-lowering "statin" drugs.

- To conduct routine liver function tests to detect hazardous side effects of treatment, such as in patients receiving hepatotoxic medications. An extension of this is therapeutic drug monitoring, in which patients receiving medications with a low therapeutic index (the ratio of the dose needed to produce a toxic effect to the dose needed to produce a therapeutic effect), such as phenytoin and carbamazepine (both of which are used to treat epilepsy), are routinely monitored for drug levels and liver function to ensure they are receiving effective and secure treatment.

6.2.2. Reference Ranges

It is crucial that the test possesses the necessary performance indicators, particularly specificity and sensitivity, for it to be regularly utilized as a clinical diagnosis aid for a given analyte in biochemical tests.

The percentage of patients with the ailment that the test successfully diagnoses is expressed as sensitivity. The percentage of healthy patients who are accurately diagnosed by the test is known as specificity.

The following mathematical formula can be used to express these two parameters:

For a given test, it would be ideal if both of these markers were 100%, but this is not always the case. This issue is most likely to arise when the variation in the test analyte concentration in the clinical sample is minimal compared to the values within the normal range.

Both of these signs describe how well the test worked, but it's also critical to be able to estimate the likelihood that a patient who received a positive test has the condition being tested for. The test's predictive ability is the best way to accomplish this.

The percentage of patients with a positive test result who are diagnosed as having the disease is expressed in this way:

The concept of predictive power can be explained using the example of fetal screening for neural tube abnormalities and Down's syndrome. The measurement of Alpha -fetoprotein (AFP), human chorionic gonadotropin (hCG), and unconjugated oestriol (uE_3) in the mother's blood serves as the basis for preliminary tests for these diseases in unborn offspring.

In comparison to the average for healthy pregnancies, the presence of these disorders causes an increase in hCG and a decrease in AFP and uE_3. The likelihood that the infant will have these disorders is estimated using the test findings along with the gestational and mother ages.

If the risk is high, additional tests are performed, such as retrieving some fetal cells from the amniotic fluid surrounding the fetus in the pregnancy for genetic screening using a hollow needle (amniocentesis).

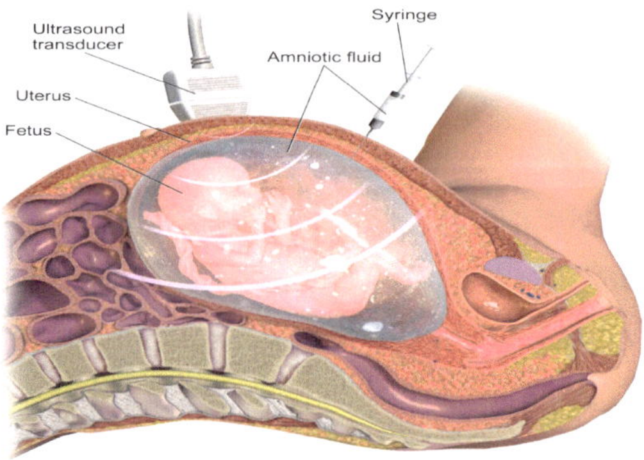

Figure 6.3. Amniocentesis.

Source: Image by Wikimedia Commons

Two out of three (67%) cases of Down's syndrome and four out of five (80%) cases of neural tube abnormalities are detected by the three tests. As a result, although the tests' performance indicators are not 100%, they are high enough to support everyday use.

The appropriate reference range to compare the test results to will have a significant impact on how accurately the biochemical test results are interpreted. In healthy individuals, the majority of biological analytes with diagnostic value are subject to significant inter- and intra-individual variation, as previously mentioned.

The analytical method selected for a given analyte assay will have its own precision, accuracy, and selectivity that will affect the analytical results. Individual laboratories must develop their reference range for each tested analyte using their chosen methodology and a significant number (hundreds) of "healthy" persons in light of these biological and analytical considerations.

The difficulty of defining "normal" and the use of invasive techniques, like venipuncture, to obtain the requisite biological samples make the recruiting of people to be included in reference range research a significant practical and ethical concern. It is particularly difficult to define reference ranges for kids, especially newborns.

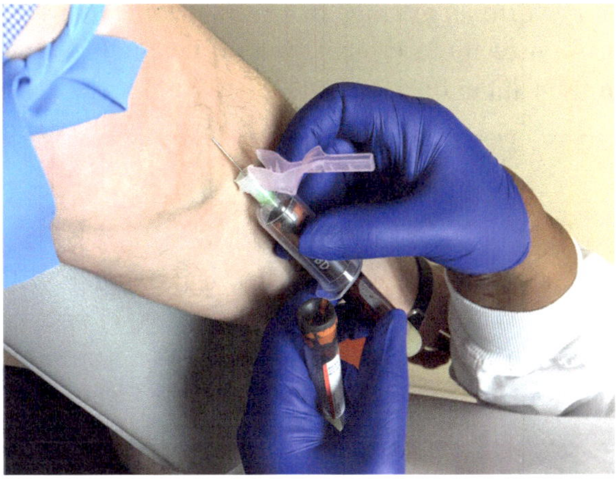

Figure 6.4. Venipuncture using a BD Vacutainer.

Source: Image by Wikimedia Commons

The range that spans the mean of the experimental population ±1.96 standard deviations is how reference ranges are most frequently expressed. 95% of the population is covered within this range.

Although most reference ranges are based on the normal distribution of individual values, there are some instances when the experimental data are asymmetric and frequently skewed to the higher limits. In such cases, it is normal to use logarithmic data to establish the reference range but even so, the range may overlap with values found in patients with the test disease state.

6.3. CLINICAL MEASUREMENTS AND QUALITY CONTROL

6.3.1. The Operation of Clinical Biochemistry Laboratories

Depending on the location in the UK, the clinical biochemistry laboratory of a typical general hospital serves a population of over 400 000 people that includes 60 General Practitioner (GP) groups.

Each weekday, this population will cause GPs and hospital doctors to request 1200 or so clinical biochemical testing on their patients. The laboratory must perform an average of seven specific analyte tests for each patient request. As a result, a typical general hospital laboratory performs 2.5 to 3 million tests annually.

> The majority of clinical biochemistry laboratories provide up to 200 distinct clinical biochemical tests, which can be broken down into eight categories, to the nearby medical community.

The majority of requests for biochemical testing will come from ordinary everyday activities, but some will come from unexpected medical emergencies that can happen at any time. The laboratory must rely significantly on automated analysis to carry out the tests and on information technology to analyze the data due to the high volume of daily test samples and the requirement for 24-hour, 7-day service.

A clinical biochemistry laboratory's three primary responsibilities are to:

- To give the requesting general practitioner (GP) or hospital doctor advice on the right tests to do for a specific medical condition as well as on how to collect, store, and transport patient samples for analysis.

- To deliver a high-quality analytical service for the fast and suitable measurement of biological analytes
- To offer a data interpretation and advice service on the results of the biochemical tests and any follow-up tests to the doctor who made the request.

The User Handbook, which was created by senior laboratory staff and includes a description of each test offered, instructions for sample collection and storage, regular laboratory operating hours, and an estimate of how long the laboratory will need to complete each test, generally supports the advice given to the clinician.

Depending on the specialty of the test, this turnaround time can range from a few hours to many weeks. Most varieties of biochemical tests are performed on serum or plasma that is obtained from blood samples.

Although the serum is the preferred matrix for biochemical tests, the amounts of the majority of test analytes in the two fluids are nearly identical. Allowing the blood to coagulate and then centrifuging the clot to retrieve the serum. The blood sample must be treated with an anticoagulant and the red blood cells must be centrifuged out in order to obtain plasma.

Heparin and EDTA are the two most popular anticoagulants, with the choice depending on the specific biochemical test needed. For instance, EDTA complexes calcium ions, making calcium invisible in EDTA plasma.

Fluoride oxalate is added to the sample for the purpose of measuring glucose, not as an anticoagulant but rather to prevent glycolysis while the sample is being transported and stored. For the storage of blood samples, specialized vacuum collecting tubes with certain anticoagulants or other additives are available.

Additionally, collection tubes with clot enhancers to hasten the clotting procedure for serum preparation are available. A gel having a specific gravity that allows it to float between cells and serum and act as a barrier between the two for up to four days is used in several containers.

Since the cellular component will have undergone lysis during the course of these 4 days, any subsequent contamination of the serum will also include intracellular components. Whole blood, urine, cerebrospinal fluid (the liquid surrounding the spinal cord and brain), feces, perspiration, saliva, and amniotic fluid can all be subjected to biochemical assays.

The collection of the samples in the proper container at the proper time, labeling with the proper patient and biohazard information, and timing are all crucial (this is especially true if the test involves measuring hormones like cortisol, which is released during the day).

To aid in the interpretation of the results and the identification of additional appropriate tests, samples for biochemical tests are submitted to the laboratory with a request form that has been signed by the requesting clinician and contains information about the tests that will be performed.

Laboratory reception

Both the sample and the request form will be given an acquisition number when they are received at the lab. This number is often written with a bar code but can also be read optically. To make sure the appropriate container has been used for the needed tests, the validity of the sample information is checked on both the request form and the sample container.

At this point, samples may be rejected if the provided information does not follow the established methodology. After selecting the appropriate samples from the request form, serum or plasma is usually prepared for analysis by centrifugation.

The tests requested by the doctor are entered into the database once the request form has been processed by the computer system, which verifies the patient's identity against the sample acquisition number. It is crucial that the sample and patient data match at this stage of reception and that the proper information is entered into the database.

Name, address, date of birth, CHI number (a unique identifier for each individual in the UK for health purposes), hospital or Accident and Emergency number, and acquisition number are some of the information that must be sufficient to specifically identify the patient given the number of potential patients in the catchment area.

6.3.2. Analytical Organization

Three work areas mainly govern the analytical organization of the majority of clinical biochemical laboratories:

- Auto-Analyzer Section
- Immunoassay Section
- Manual Section

Auto-Analyzer Section:

Numerous commercial vendors offer auto-analyzers specifically designed for clinical biochemical analysis.

The majority of analyzers are fully automated and equipped with carousels for holding test samples in racks that can hold up to 15 samples each, one or two carousels for up to 60 different reagents, each of which is identified by a different bar code, carousels for washing and preparing samples, and a reaction carousel with up to 200 cuvettes for starting and monitoring individual test reactions.

Three techniques are used by the Abbott ARCHITECT c1600:

- Spectrophotometry
- Immunoturbidimetry
- Potentiometry

The potentiometric system, based on the use of ion-selective electrodes (ISEs) that are combined into a single unit, is used to measure electrolyte concentrations of Na^+, K^+, and Cl^- simultaneously with an analytical time of less than 4 min on a sample size of only 25 mm^3. The spectrophotometry system can measure at 16 wavelengths simultaneously.

330 cuvettes are present on the reaction carousel, and an eight-stage wash system uses automatic cleaning and drying to clean and dry each cuvette before each assay. When necessary, additional washes are offered to lessen carryover. Testing is arranged into two parallel lines, each of which is a resource, and analysis-controlled separately, to maximize throughput.

Reagents for up to 62 tests based on various analytical techniques may be loaded at the same time. The cycle time of the analyzer is fixed. Readings are taken every entire rotation of the reaction carousel, which rotates about a quarter of a turn every 4.5 seconds.

The assay is run and the assay parameters affect how long the analysis will take overall. All tests take less than ten minutes to complete, and the results are presented right away. Complete findings for a patient sample are recorded as soon as the last essay is completed.

The analyzer can theoretically process 1800 tests per hour. The c1600 is an "open system," like the majority of analyzers, in that lab staff can add "in-house" tests to the routine in addition to the programmed tests. These could include particular protein assays, drug misuse diagnostics, and therapeutic medication monitoring.

In order for one analyzer to support the other, each laboratory will have at least two analyzers, each of which offers a comparable analytical repertoire. Each sample's bar code acquisition number is read by the analyzer, which then queries the host computer database to determine which tests should be run on the sample based on the reading.

The analyzer is programmed to take the appropriate volume of a sample using a sampler that may also be able to detect micro clots in the sample, add the appropriate volume of reagents in a specified order, and monitor the progress of the reaction after the identified tests are automatically prioritized into the most effective order.

The same methodology is used to analyze internal quality control samples regularly. All reagents are automatically tracked by the analyzer so it can determine when they need to be replenished.

The operator has the option of validating the test findings either on the analyzer or on the main computer database once they have been calculated. When appropriate, the outcomes might also be compared to earlier outcomes for the same patient.

Micro Sensor Analyzers:

The creation of miniaturized multi-analyte sensor blocks, which are frequently used in clinical biochemistry laboratories for the routine measurement of pH, pCO_2, Na^+, K^+, $Ca2^+$, Cl^-, glucose, and lactate, was sparked by recent advancements in microsensor technology.

A "cartridge" that includes the analyte sensors, a reference electrode, a flow system, wash solution, waste receptacle, and a process controller is the block's fundamental component. The block's surface acts as an interface for the analyte sensors and is thermostatically maintained at 37 degrees Celsius. The sensors are embedded in three layers of plastic, the size, and shape of a credit card.

There can be up to 24 sensors on each card. Under each sensor, a metallic contact creates an electrical connection to the cartridge. The processes unique to each analyte produce a current as the test sample travels over the sensors, which is then recorded.

The analyte concentration in the fluid in the sample passage directly relates to the current's magnitude. The concentration of the test sample can be determined by calibrating the sensors using standard analyte solutions. After being automatically cleaned after each test sample, the sensor card and sample path can be utilized to analyze up to 750 whole blood samples before being discarded.

A platinum amperometric electrode with a positive potential in relation to the reference electrode is used in the glucose and lactate sensors.

The glucose oxidase reacts with the glucose and oxygen in the measurement of glucose to produce hydrogen peroxide, which then diffuses through a regulating layer and is oxidized by the platinum electrode to release electrons and produce a current flow, the size of which is proportional to the rate of hydrogen peroxide diffusion.

The GEM Premier 4000 system has an active quality control process controller that keeps an eye on the system's performance, verifies the cartridge's integrity, and tracks the electrode's reaction to detecting micro clots in the test sample that could invalidate the analytical results.

Immunoassay Section

Modern auto-analyzers perform immunoassay processes mostly using fluorescence or polarized fluorescence techniques. The variety of analytes varies from manufacturer to manufacturer but typically includes drugs of abuse, medicinal medications (such as theophylline and digoxin), and basic endocrinology (such as thyroid function tests) (opiates, cannabis).

When used in immunoassay mode, auto-analyzers operate similarly to those previously mentioned. Results are typically reported on the same day and are typically compared to the patient's prior set of results.

Manual assays section

This method of conducting biochemical tests is typically more labor-intensive than the other two categories.

It covers a variety of analytical techniques, including immunoelectrophoresis, acetate or gel electrophoresis, and some more challenging basic spectrophotometric assays. 5-hydroxyindole acetic acid (for the diagnosis of carcinoid syndrome), assays for catecholamines (for the diagnosis of phaeochromocytoma), or HbA1c (for the monitoring of diabetes) are some examples.

Result reporting

0Analytical results are initially validated by the section leader or the instrument operator. Internal quality control processes for certain analytes will be used in this validation process, in part. Each batch of test analytes includes a quality control sample that is examined at least twice per day.

The analytical data are then put through an automatic process that detects results that are either significantly aberrant or need clinical comment or interpretation in accordance with guidelines established by senior laboratory staff.

6.3.3. Neonatal Screening

Newborn screening involves examining newborns for several potentially harmful conditions. Early detection of these diseases allows for the adoption of preventative measures that assist shield the youngster from the disorders. However, such testing is not easy due to the difficulty of obtaining adequate samples of biological fluids for the tests. The creation of tandem MS methods has greatly reduced this issue. A single dried blood spot measuring 3 mm in diameter can be used as a screening tool for a variety of metabolic illnesses. The approach is just utilized to send the material to the MS; pre-HPLC separation of the sample is not required.

The method can be used to screen for a wide range of inherited metabolic diseases, including aminoacidopathies like phenylketonuria (PKU), which is caused by a deficiency of the enzyme phenylalanine hydroxylase, fatty acid oxidation defects like medium-chain acyl-CoA dehydrogenase deficiency (MCADD), and organic acidemias like propionic acidemia, which is caused by a deficiency of the enzyme propionyl-CoA carboxylase.

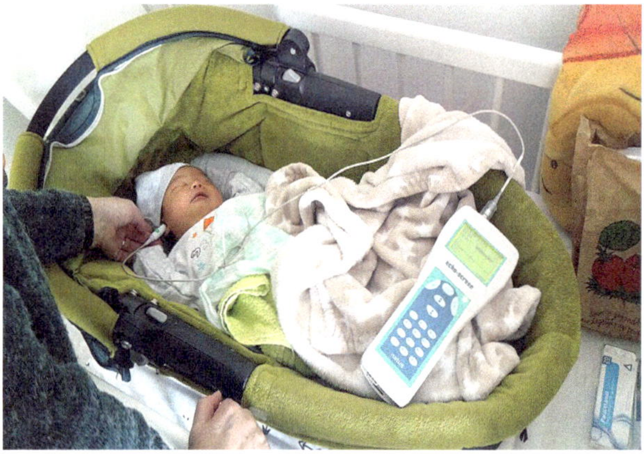

Figure 6.5. Newborn hearing screening.

Source: Image by Wikimedia Commons

6.3.4. Quality Assessment Procedures

A clinical biochemistry department will participate in external quality assessment programs in addition to regularly carrying out internal quality control procedures that involve the repeated analysis of reference samples covering the full analytical range for the test analyte in order to validate the analytical precision and accuracy of the biochemical tests carried out by the department.

There are two primary national clinical biochemistry external quality assessment schemes in the UK: the Wales External Quality Assessment Scheme (WEQAS: www.weqas.com), coordinated at the University Hospital of Wales, Cardiff, and the UK National External Quality Assessment Scheme (UK NEQAS: www.ukneqas.org.uk) coordinated at the Queen Elizabeth Medical Centre.

The majority of clinical biochemistry departments at UK hospitals participate in both schemes. Every two weeks, test samples based on human serum are distributed by UK NEQAS and WEQAS. The samples used for UK NEQAS contain a number of analytes, each at an undisclosed concentration that falls within the analytical range. Each analyte's concentration varies from one distribution to the next.

While WEQAS disperses four or five test samples, each of which contains the test analytes at various concentrations that fall within the analytical range, For example, general chemistry analytes, peptide hormones, steroid hormones, and therapeutic drug monitoring analytes are only a few examples of the linked analytes found in groups of related test samples provided under the UK NEQAS and WEQAS quality assessment programs. Participating laboratories decide to sign up for plans related to their analytic services.

Along with routine clinical samples, the participating laboratories are expected to examine samples for external quality assessments and report the findings to the coordinating center. Each center conducts a thorough statistical analysis of all the results submitted and provides a confidential report to the individual laboratories.

The statistical data are recorded and then broken down into different methods (such as the glucose oxidase and hexokinase methods for glucose) and for particular manufacturers' systems. They compare the data from each laboratory with all the submitted data as well as with the combined data.

Results are shown in tabular, graphical, and histogram form, and they are contrasted with the outcomes of recently submitted samples. Longer-term trends in analytical performance for each analyte to be monitored are made possible by this comparison with prior performance data.

6.3.5. Clinical Audit and Accreditation

Clinical laboratories are subject to clinical audits in addition to taking part in external quality assessment programs. This is a methodical and critical evaluation of the laboratory's overall performance in comparison to both its own proclaimed standards and practices and to standards that have been adopted nationally.

When it comes to analytical methods, the audit assesses the laboratory's performance in terms of the appropriateness of using the tests the laboratory offers, the clinical interpretation of the findings, and the operational processes for the receiving, analysis, and reporting of test samples.

To implement change and improve, the audit is therefore primarily focused on the procedures that produce test data, even though it also includes the examination of analytical data. The audit's main goal is to make sure the patient gets the finest support and care while remaining as cost-effective as feasible.

The audit often lasts a few days and involves interaction with all laboratory workers. It is conducted by junior doctors, laboratory staff, or hospital-based CPA assessors.

Accreditation is a process that is closely related to clinical audits. Clinical audit, on the other hand, is done primarily for the local benefit of the laboratory and its employees, and ultimately for the benefit of the patient, whereas accreditation is a national and public acknowledgment of the professional standing and quality of the laboratory and its staff.

Either a government department or agency or a recognized public professional association is in charge of the accreditation procedure and assessment. Different models are used in various nations.

Government agencies in the UK demand that clinical biochemistry laboratories be accredited, which is done by either Clinical Pathology Accreditation (UK) Ltd (CPA) or, less frequently, the United Kingdom Accreditation Service (UKAS). In the USA, accreditation is required and may be administered by one of several "deemed authorities," including the College of American Pathologists. Some organizations accredit non-clinical analytical laboratories as well. Examples in the UK include the British

Standards Institution (BSI) and the National Measurement Accreditation Service (NAMAS).

Three of the numerous international forums for the harmonization of national accreditation standards for analytical laboratories are the International Accreditation Co-operation (ILAC), the European Co-operation for Accreditation (EA), and the Asia-Pacific Laboratory Accreditation Co-operation (APLAC).

The conformity of the laboratory with the standards defined by the accreditation body is evaluated by assessors chosen by the accreditation body.

The standards address a wide range of topics, including precision and accuracy, timeliness of results, the clinical relevance of the tests performed, competence to perform the tests as determined by the training and qualifications of the laboratory staff, health and safety, the caliber of administrative and technical support systems, and the caliber of laboratory management systems and document control. National recognition is the result of an assessment that is successful.

6.4. EXAMPLES OF BIOCHEMICAL AIDS TO CLINICAL DIAGNOSIS

6.4.1. Principles of Diagnostic Enzymology

It has long been known that measuring the activity of large quantities of specific enzymes in serum can help with clinical diagnosis and prognosis. Based on where they perform their physiological functions normally, the enzymes found in serum can be categorized into three groups:

- ***Serum-specific enzymes***: These enzymes rely on serum for their regular physiological function. Examples include the enzymes involved in blood coagulation and lipoprotein metabolism.
- ***Secreted enzymes***: Serum-specific enzymes are closely connected to secreted enzymes. Pancreatic lipase, prostatic acid phosphatase, and salivary amylase are a few examples.
- ***Non-serum-specific enzymes***: These enzymes play no physiological part in serum. As a result of regular cell turnover or more frequently as a result of cell membrane injury, cell death, or morphological alterations to cells, such as those in cases of malignancy, they are released into the extracellular fluid and

subsequently show up in serum. The serum's typical cofactors and/or substrates may be deficient or nonexistent.

The greatest diagnostic value can be found in the third group of serum enzymes. When a cell is broken, its contents are released over the course of several hours, with cytoplasmic enzymes showing first because their release is solely dependent on the integrity of the plasma membrane being compromised.

These enzymes have a huge concentration gradient across the membrane, more than a thousand-fold, which makes it easier for them to release themselves after cell membrane injury. The integrity of the cell membrane is especially vulnerable to situations that hinder energy generation, such as a limitation in the oxygen supply.

It is also susceptible to harmful substances, such as some medications, bacteria, immunological disorders, and genetic flaws. Such events may not always result in the same relative numbers of enzymes being released from cells into the serum as were initially present in the cell.

These variations reflect variances in their serum half-lives as well as differences in the rate of their metabolism and excretion from the body. Glutathione S-transferase, creatine kinase and intestinal alkaline phosphatase are a few-hour to several-day tests (liver alkaline phosphatase, alanine aminotransferase, lactate dehydrogenase).

The clinical exploitation of non-serum-specific enzyme activities is influenced by several factors:

- ***Organ specificity***: Few enzymes are exclusive to a single organ, but thankfully some enzymes are found in some tissues in much higher concentrations than in others. Therefore, a number of enzymes identified in serum have relative proportions (patterns) that are frequently indicative of the organ of origin.
- ***Isoenzymes***: Some clinically significant enzymes can be found in isoenzyme forms, and analysis of the serum isoenzymes can reveal the organ of origin in many cases when the relative proportion of the isoenzymes varies significantly between tissues.
- ***Reference ranges***: The activity of enzymes seen in serum from healthy people is almost always lower than those found in serum from those who have a clinical condition such as liver disease. In many instances, the degree of cellular damage to the organ of origin can be determined directly by the degree to which a certain enzyme's activity is increased by the disease state.

- ***Variable rate of increase in serum activity***: Each enzyme has a unique feature, which varies depending on how quickly the activity of released enzymes in serum increases after cell injury in a specific organ. Additionally, a useful indicator of the patient's recovery from the condition is the pace at which each enzyme's activity falls towards the reference range after the incident that produced cell damage and the patient's subsequent treatment.

The use of diagnostic enzymology in the treatment of heart disease and liver illness serves as an example of the practical application of the many points made regarding its uses.

6.4.2. Ischaemic Heart Disease and Myocardial Infarction

The presence of oxygen is necessary for the heart to operate properly. The gradual buildup of cholesterol-rich atheromatous plaques in the coronary arteries may jeopardize this oxygen availability. As these deposits grow, a point is reached where the oxygen supply is insufficient during periods of peak demand, such as during periods of intense activity.

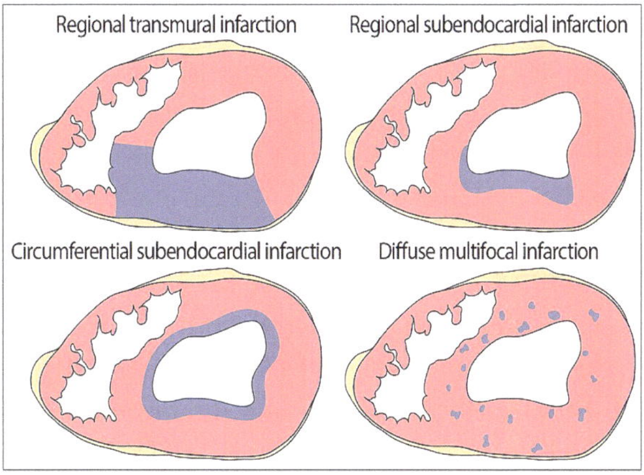

Figure 6.6. Patterns of topographic distribution of myocardial infarction.

Source: Image by Wikimedia Commons

As a result, the person has intense chest pain, often known as angina pectoris ('angina of effort'), and their heart briefly becomes temporarily ischaemic ('lacking in oxygen'). The cardiac cells are not harmed and do not leak their internal contents, despite the fact that the pain may be very intense during such episodes.

The patient will nonetheless experience a myocardial infarction (also known as a "heart attack") if the arteries are completely blocked, either by the plaque or by a small thrombus (clot) that is prevented from passing through the artery by the plaque.

This condition is characterized by the same severe chest pain, but in this instance, the pain is also accompanied by irreversible damage to the cardiac cells and the release of their cellular contents. This emission happens gradually over several hours rather than immediately.

The clinician to determine whether or not myocardial infarction was present along with the chest discomfort from the perspective of clinically managing the patient. The patient does not experience the typical chest pain in around one-fifth of myocardial infarction instances (this is known as a "silent myocardial infarction"), but it is still crucial for the clinician to be aware that the event has occurred.

Electrocardiogram (ECG) patterns are a key signal of these occurrences, but in atypical presentations, ECG changes can be unclear, necessitating the search for further information in the form of variations in blood enzyme activity. Three enzymes whose activity are frequently assessed include:

- **Creatine kinase (CK):** This enzyme converts phosphocreatine, which is necessary for proper muscle metabolism, into creatine. Three isoforms of the protein CK—CK-MM, CK-MB, and CK-BB exist because it is a dimeric protein made up of two monomers, one of which is designated as M (for muscle) and the other as B (for the brain).

Since these isoenzymes are distributed in different tissues, the cardiac muscle contains 80–85 percent MM and 15–20 percent MB, skeletal muscle contains 99% MM and further 1% MB, and the brain, stomach, colon, and bladder predominately contain BB.

Even when the total CK activity is still within the reference range, the CK-MB form's elevated serum activity, which is virtually exclusively found in the heart, provides clear evidence of myocardial infarction. CK activity is elevated in several clinical situations.

Within six hours after myocardial infarction, there is a spike in total serum CK activity that may be seen, and it peaks 24 to 36 hours later. A rise in CK-MB, on the other hand, can be detected within 3–4 hours, reaches 100% sensitivity within 8–12 hours, and reaches a peak within 10–24 hours. For 2-4 days, it is elevated.

- **Aspartate aminotransferase (AST):** This particular transaminase is one of many in intermediate metabolism. The majority of tissues contain it; however, the heart and liver contain a lot of it.

Following myocardial infarction, its serum activity is increased and peaks between 48 and 60 hours later. Although it has limited clinical benefit in the early identification of heart muscle damage, it can be helpful when chest pain presents later than expected.

- **Lactate dehydrogenase (LD):** There are five different isoforms of this tetrameric protein, which have two monomers indicated by the letters H (heart) and M (muscle): LD-1 (H4), LD-2 (H3M), LD-3 (H2M2), LD-4 (HM3), and LD-5 (M4). In the heart, brain, and kidney, LD-1 predominates, while LD-5 predominates in the skeletal muscle and liver.

Following myocardial infarction, total LD activity and LD-1 activity in serum rise and peak 48–72 hours later. As a result, the activity declines considerably more gradually than it does for CK or AST. Monitoring the patient's recovery from the myocardial infarction occurrence is primarily the extent of the diagnostic significance of LD activity measurement.

6.4.3. Liver Disease

Diagnostic enzymology is routinely used to discriminate between several forms of liver disease including:

- **Hepatitis**: The most frequent cause of general liver inflammation is a viral infection, but it can also result from glandular fever or blood poisoning (septicemia). There is just a minor release of cellular enzymes as a result of the moderate necrosis of the liver cells.
- **Cirrhosis**: Hepatic cells are generally destroyed and replaced by fibrous tissue. It is most frequently brought on by excessive alcohol consumption, although it can also be brought on by chronic hepatitis, a number of autoimmune diseases, and hereditary disorders. They all cause significant cell destruction and the release of enzymes from liver cells.
- **Malignancy**: Primary and secondary tumors.
- **Cholestasis**: Preventing bile from entering the gut due to either bile duct blockage from gallstones or tumors or liver cell death

from cirrhosis or persistent hepatitis. Because of this, obstructive jaundice develops (presence of bilirubin, a yellow metabolite of haem, in the skin).

Patients with these different liver diseases frequently present to their doctor with similar symptoms, so a differential diagnosis must be made using a variety of investigations, including imaging techniques, particularly ultrasonography (ultrasound), magnetic resonance imaging (MRI), computerized tomography (CT) scanning, microscopic examination of biopsy samples, and liver function tests.

Typically, four enzymes are measured to aid in differential diagnosis:

- **Aspartate aminotransferase (AST) and alanine aminotransferase (ALT):** Although these enzymes are generally present, their ratios in serum are indicative of the precise source of liver cell destruction, as was previously mentioned. Acute viral hepatitis and fresh obstructive jaundice, for instance, are characterized by an AST/ALT ratio of less than 1, obstructive jaundice brought on by viral hepatitis is characterized by an AST/ALT ratio of approximately 1, and cases of cirrhosis are characterized by an AST/ALT ratio of greater than 1.
- **g-Glutamyl transferase (GGT):** This enzyme can be measured by using the substrate g-glutamyl-4-nitroaniline and watching for the release of 4-nitroaniline at a wavelength of 400 nm. It transfers a g-glutamyl group between substrates. The liver, particularly the bile canaliculi, kidney, pancreas, and prostate all contain large amounts of GGT, however, these organs do not manifest themselves by raising serum levels. Increased activities are seen in cirrhosis, secondary hepatic tumors, and cholestasis and frequently co-occur with an increase in alkaline phosphatase activity, particularly in the latter. Alcohol stimulates its production, and several medicines also increase its serum activity.
- **Alkaline phosphatase (AP):** Although this enzyme can be found in most tissues, it is particularly prevalent in the placenta, kidney, bone, and bile canaliculi. It might be tested by releasing 4-nitrophenol at 400 nm while utilizing 4-nitrophenyl phosphate as a substrate. Its activity is increased in obstructive jaundice, and when it is tested alongside ALT, it can be used to differentiate between obstructive jaundice and hepatitis. Its activity is increased more than ALT's in obstructive jaundice.

Decreasing serum activity of AP is valuable in confirming an end of cholestasis. Raised serum AP levels can also be present in various bone diseases and during growth and pregnancy.

6.4.4. Kidney Disease

The liver and kidneys are the two principal organs in charge of eliminating waste from the body. The management of electrolyte and water balance, as well as the production of erythropoietin, are additional distinct roles performed by the kidneys. About 1 million nephrons, which receive blood traveling to the kidneys, are present in each of the two kidneys. The glomerulus of each nephron, which filters the plasma water to form the ultrafiltrate or primary urine, receives blood before it reaches the kidneys.

This process removes all of the plasma's constituents other than proteins, leaving only water in the ultrafiltrate. In a healthy adult, the two kidneys produce about 100-140 cm^3 of primary urine every minute, or 200 dm^3 per day, with each nephron producing approximately 100 mm^3 of primary urine per day. The term for this is "glomerular filtration rate" (GFR).

The primary urine next enters the nephron's tubule, which is where water, lipophilic chemicals, and cellular nutrients like carbohydrates and amino acids are reabsorbed both actively and passively as well as secreted actively in some cases. Approximately 2 dm^3 of urine is produced each day as a result of these two processes working together and stored in the urinary bladder.

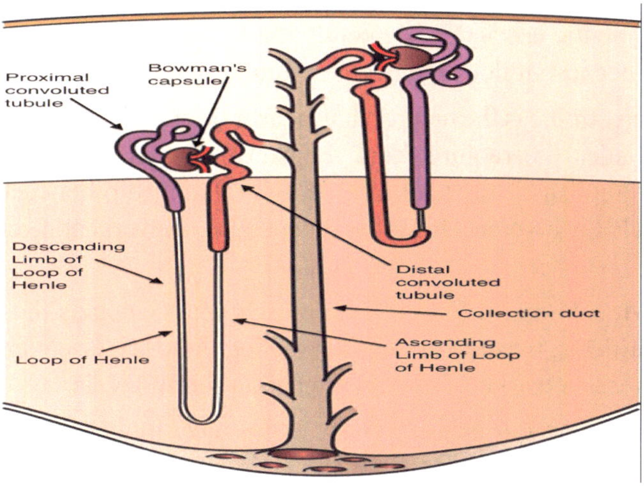

Figure 6.7. A kidney nephron and its structure.

Source: Image by Wikimedia Commons

The most accurate measure of kidney function is currently available in the glomerular filtration rate. Reduced GFR, which is a sign of any renal pathology, has substantial physiological repercussions including anemia and severe cardiovascular disease.

Glomerular nephritis is an example of a subacute or intrinsic renal disease that progresses into chronic kidney disease (CKD) from kidney disease. Dialysis and kidney transplants are required in cases of total renal failure.

There is proof that the prevalence of CKD is rising in developed nations and is linked to a higher risk of diabetes and an increasingly elder population. Therefore, there is a significant clinical need for precise GFR measurements to identify the development of renal disease, evaluate how severe it is, and track its subsequent progression.

Measurement of Glomerular Filtration Rate

The idea of renal clearance, which is defined as the volume of serum cleared of a specific substance by glomerular filtration in unit time, is the foundation for the measurement of GFR. As a result, it has units for $cm^3\ min^{-1}$.

In theory, the measurement might be based on any endogenous or exogenous chemical that is exposed to glomerular filtration and is not reabsorbed. The polysaccharide inulin satisfies these requirements and is not subject to many variations or interferences, but it is not convenient for normal therapeutic use and is frequently used as a benchmark for alternative approaches because it does not occur naturally in the body.

Serum creatinine is the marker that is most frequently utilized in practice. Since it satisfies the requirements for excretion and is the byproduct of creatine metabolism in skeletal muscle, the relationship between its serum levels and GFR is inverse. However, it depends on several non-renal factors, such as:

- **Muscle mass**: Extremes in muscle mass, such as those found in athletes, those with muscle-wasting diseases, or patients who are undernourished, have an impact on serum levels.
- **Gender**: For a given GFR, males have higher serum creatinine levels than females.
- **Age**: Serum creatinine levels are lower in children under 18 and higher in the elderly.

- **Ethnicity**: In comparison to Caucasians, African Caribbeans have a greater serum creatinine for a given GFR.
- **Drugs**: Some regularly used medications, including cimetidine, trimethoprim, and cephalosporins, interfere with the excretion of creatinine and raise GFR readings as a result.
- **Diet**: Red meat and oily fish consumption recently can increase serum creatinine levels. Regular laboratory calculations of GFR (referred to as eGFR) are based on measuring blood creatinine concentration and eGFR is calculated from it using an algorithm that corrects for the four aforementioned variables. One of two methods is frequently used to measure serum creatinine:
- **Spectrophotometric method based on the Jaffe reaction**: Alkaline picric acid reagent is used in this process to create a red product that is visible at 510 nm. One drawback is that the reagent also reacts with several non-creatinine chromogens, such as ketones, ascorbic acid, and cephalosporins, and as a result, produces high results.

6.4.5. Clinical Assessment of Renal Disease

Acute renal failure (ARF): A rise in serum creatinine and urea levels indicates the presence of acute renal failure, which is the failure of renal function over a period of hours or days. It is a condition that can be fatal that is brought on by the retention of salts like sodium and potassium and nitrogenous waste products.

An increased risk of cardiac arrest and ECG abnormalities may result from the surge in potassium. There are three types of acute renal failure: pre-renal, renal, and post-renal. Early detection of pre-or post-renal variables and prompt therapy implementation may enable correction before kidney damage starts.

Lack of renal perfusion results in pre-renal failure. This can happen as a result of a decrease in cardiac output brought on by cardiogenic shock, massive pulmonary embolism, cardiac tamponade (application of pressure), or other causes of hypertension like sepsis. It can also happen as a result of volume loss from hemorrhage, gastrointestinal fluid loss, and burns.

Among the post-renal reasons include bilateral uretic obstruction brought on by calculi or tumors, decreased bladder output/urethral obstruction, such as urethral stricture, or enlarged prostate brought on by hypertrophy of cancer.

Damage to the kidneys can be prevented by fixing the underlying issue. Glomerular nephritis, vascular disease, severe hypertension, hypercalcemia, invasive diseases like sarcoidosis or lymphoma, nephrotoxins such as heavy metals, aminoglycoside antibiotics, and non-steroidal anti-inflammatory medicines can all lead to abrupt renal failure.

Chronic kidney disease (CKD): A falling eGFR is a hallmark of CKD, a disorder that progresses over time. All CKD patients undergo routine clinical and laboratory evaluations, as well as further clinical care once Stage 3 has been reached. By using pharmacological therapy, the disease is being attempted to be reversed or arrested, sparing the patient the discomfort and the funding source the expense of dialysis or transplantation.

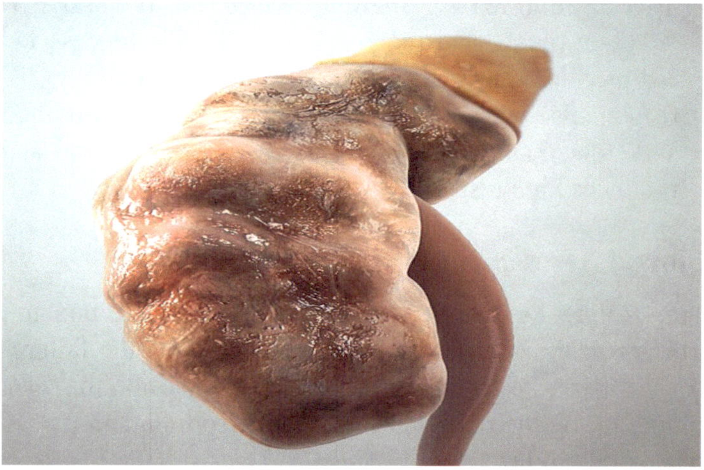

Figure 6.8. A graphic representation of a chronically affected kidney.

Source: Source: Image by Wikimedia Commons

6.5. CONCLUSION

This chapter discussed the various principles of clinical biochemistry. It also discussed the different principles of clinical biochemical analysis. In this chapter, clinical measurements and quality control such as the operation of clinical biochemistry, analytical organization, neonatal screening, etc., have been discussed. Towards the end of the chapter, it discussed the example of biochemical aids in clinical diagnosis of ischaemic heart disease, liver disease, and kidney disease.

REFERENCES

1. Fairbrother, A., 2020. Clinical Biochemistry. *Nondestructive Biomarkers in Vertebrates*, [online] pp.63-89. Available at: <https://www.taylorfrancis.com/chapters/edit/10.1201/9780367813703-5/clinical-biochemistry-anne-fairbrother> [Accessed 4 July 2022].
2. Jain, B., Goswami, S. and Pandey, S., 2021. Clinical Biochemistry. *Protocols in Biochemistry and Clinical Biochemistry*, [online] pp.101-118. Available at: <https://www.sciencedirect.com/science/article/pii/B9780128220078000015?via%3Dihub> [Accessed 4 July 2022].
3. Nexø, E., 1995. Clinical biochemistry. *FEBS Letters*, [online] 375(3), pp.312-313. Available at: <https://febs.onlinelibrary.wiley.com/doi/abs/10.1016/0014-5793%2895%2990102-7> [Accessed 4 July 2022].
4. Rumsby, G., n.d. Principles of Clinical Biochemistry. *Wilson and Walker's Principles and Techniques of Biochemistry and Molecular Biology*, [online] pp.346-380. Available at: <https://www.cambridge.org/highereducation/books/wilson-and-walkers-principles-and-techniques-of-biochemistry-and-molecular-biology/2159004E019DDD87C0A97EE8DB72B79F?chapterId=CBO9781316677056A11-1#contents> [Accessed 4 July 2022].
5. White, S., 2011. Principles and techniques of biochemistry and molecular biology, seventh edition. *Biochemistry and Molecular Biology Education*, [online] 39(3), pp.244-245. Available at: <https://iubmb.onlinelibrary.wiley.com/doi/10.1002/bmb.20499> [Accessed 4 July 2022].

CHAPTER 7

SPECTROSCOPY TECHNIQUES IN BIOCHEMISTRY

CONTENTS

As light interacts with matter, spectroscopic techniques can probe certain features of a sample to determine its structure or consistency. Different molecular features can be probed by electromagnetic radiation, which exhibits different energies. This chapter discusses the fundamental interaction of electromagnetic radiation with matter.

7.1. INTRODUCTION

The application of electromagnetic radiation cannot be divided logically into parts that are treated separately. Based on 'common practice', this text presents a rationale for splitting up the text into two parts. Molecular consistency and conformational structure of biological molecules are investigated with visible or UV light in this chapter. Biochemical scientists typically use these methods to conduct their first analyses.

When electromagnetic radiation is characterized by its properties and its interaction with matter, it becomes possible to understand the variety of types of spectra. This will enable us to understand the different spectroscopic techniques and how they can help solve biological problems.

Molecular exclusion chromatography, mass spectrometry, analysis centrifugation and electron microscopy are all techniques used to obtain clues about the structures of biomolecules and their larger assemblies. The spectroscopic techniques are discussed in further complementary methods, and by assembling the jigsaw of pieces of information, one can gain a comprehensive picture of the structure of the biological object under study.

In addition, the spectroscopic principles established are often employed as a read-out in a huge variety of biochemical assays, and several more sophisticated technologies employ these basic principles in a 'hidden' way.

Aspects of molecular interactions can be studied using spectroscopic methods. The nuclear magnetic resonance (NMR) method is one of these techniques. An NMR probe or an NMR diffusion measurement can also be used to study the interaction between mucus and polymers and/or carbon. This method does not require pretreatment or labeling.

The principle behind it is that the affinity of mucin for small, high-affinity probes such as ketotifen fumarate is highly dependent on polymers' interactions with mucus. The mucoadhesive property can be measured by comparing the affinity shift between mucin and the probe when mucoadhesive polymers are present or absent.

Researchers from Uccello-Barretta et al. determined that selective proton relaxation rates were highly responsive to this type of interaction, and they used this technique to investigate the mucoadhesive properties of tamarind seed polysaccharide, arabinogalactan, and a mixture of tamarind seed polysaccharide and hyaluronic acid with bovine submaxillary mucin.

Nonionic polymers with noncovalent bonding are suited to this technique. Many spectroscopic techniques are useful for observing polymer chain conformations. Ultraviolet spectroscopy (UV) is used almost exclusively for biopolymers to establish the degree of helicity and to study other aspects of protein and nucleic acid tertiary structure.

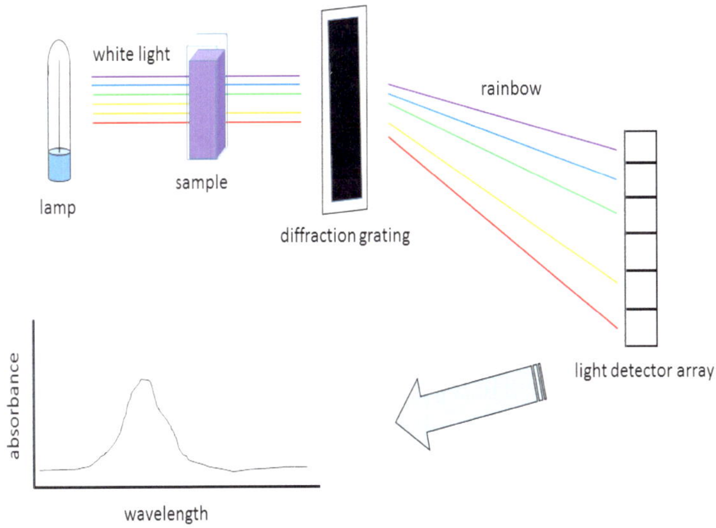

Figure 7.1. UV vis spectroscopy.

Sources: www.commons.wikimedia.org

For example, the UV spectra of nucleic acids and nucleotides show hypochromic effects when helical conformations are formed. Concerning synthetic polymers, there is some UV evidence that isotactic polystyrene partially retains a helical conformation in solution. This is confirmed by infrared measurements.

Spectroscopic techniques have been applied in virtually all technical fields of science and technology. Radio-frequency spectroscopy of nuclei in a magnetic field has been employed in a medical technique called magnetic resonance imaging (MRI) to visualize the internal soft tissue of the body with unprecedented resolution.

Microwave spectroscopy was used to discover the so-called three-degree blackbody radiation, the remnant of the big bang (i.e., the primeval explosion) from which the universe is thought to have originated. The internal structure of the proton and neutron and the state of the early universe up to the first thousandth of a second of its existence are being unraveled with spectroscopic techniques using high-energy particle accelerators.

The constituents of distant stars, intergalactic molecules, and even the primordial abundance of the elements before the formation of the first stars can be determined by optical, radio, and X-ray spectroscopy. Optical spectroscopy is used routinely to identify the chemical composition of matter and to determine its physical structure.

7.2. PROPERTIES OF ELECTROMAGNETIC RADIATION

The interaction of electromagnetic radiation with matter is a quantum phenomenon and is dependent upon both the properties of the radiation and the appropriate structural parts of the samples involved. This is not surprising, since the origin of electromagnetic radiation is due to energy changes within matter itself. The transitions which occur within matter are quantum phenomena and the spectra which arise from such transitions are principally predictable.

Electromagnetic radiation is composed of an electric and a perpendicular magnetic vector, each one oscillating in a plane at right angles to the direction of propagation. The wavelength Λ is the spatial distance between two consecutive peaks (one cycle) in the sinusoidal waveform and is measured in submultiples of meters, usually in nanometers (nm). The maximum length of the vector is called the amplitude.

The frequency υ of the electromagnetic radiation is the number of oscillations made by the wave within the timeframe of 1 s. It, therefore, has u of 1 s^{-1}=1 Hz. The frequency is related to the wavelength via the speed of light c (c = 2.998 10^8 ms^{-1} in vacuo) by $\upsilon = c\ \Lambda^{-1}$. A historical parameter in this context is the wavenumber υ which describes the number of completed wave cycles per distance and is typically measured in 1 cm.

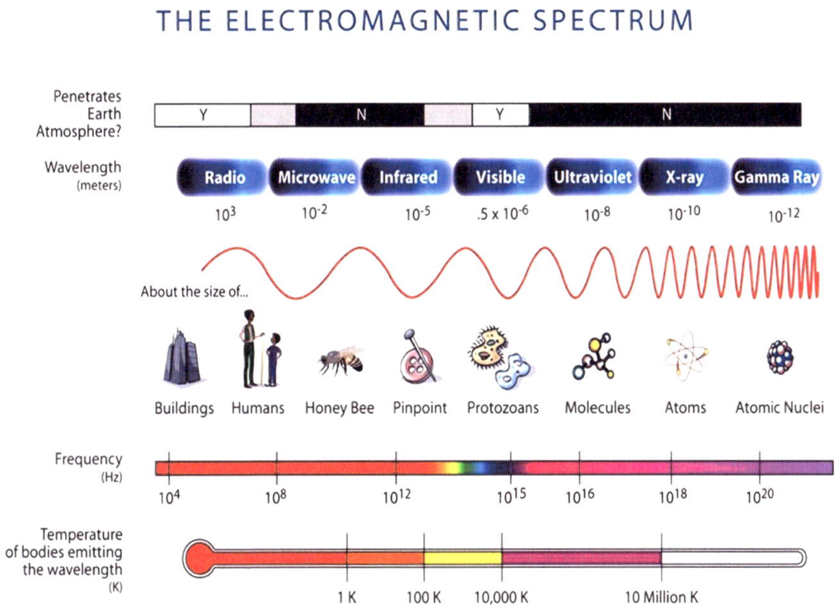

Figure 7.2. Electromagnetic spectrum.

Sources: www.flickr.com

7.3. INTERACTION WITH MATTER

The spectrum of electromagnetic radiation is organized by increasing wavelength, and thus decreasing energy, from left to right. Also annotated are the types of radiation and the various interactions with matter and the resulting spectroscopic applications, as well as the interdependent parameters of frequency and wavenumber.

Electromagnetic phenomena are explained in terms of quantum mechanics. The photon is the elementary particle responsible for electromagnetic phenomena. It carries electromagnetic radiation and has properties of a wave, as well as of a particle, albeit having a mass of zero. As a particle, it interacts with matter by transferring its energy E:

$$E = \frac{hc}{\lambda} = h\nu \qquad \textbf{EQ. 7.1}$$

where h is the Planck constant ($h = 6.63\ 10^{-34}$ Js) and v is the frequency of the radiation.

When considering a diatomic molecule, rotational and vibrational levels possess discrete energies that only merge into a continuum at very high energy. Each electronic state of a molecule possesses its own set of rotational and vibrational levels. Since the kind of schematics is rather complex, the Jablonski diagram is used instead, where electronic and vibrational states are schematically drawn as horizontal lines, and vertical lines depict possible transitions.

For a transition to occur in the system, energy must be absorbed. The energy change ΔE needed is defined in quantum terms by the difference in absolute energies between the final and the starting state as $\Delta E = E_{final} - E_{start} = hv$.

Electrons in either atoms or molecules may be distributed between several energy levels but principally reside in the lowest levels (ground state). For an electron to be promoted to a higher level (excited state), energy must be put into the system. If this energy $E = hv$. is derived from electromagnetic radiation, this gives rise to an absorption spectrum, and an electron is transferred from the electronic ground state (S_0) into the first electronic excited state (S_1).

The molecule will also be in an excited vibrational and rotational state. Subsequent relaxation of the molecule into the vibrational ground state of the first electronic excited state will occur. The electron can then revert to the electronic ground state. For non-fluorescent molecules, this is accompanied by the emission of heat (ΔH).

The plot of absorption probability against wavelength is called absorption spectrum. In the simpler case of single atoms (as opposed to multi-atom molecules), electronic transitions lead to the occurrence of line spectra. Because of the existence of more different kinds of energy levels, molecular spectra are usually observed as band spectra which are molecule-specific due to the unique vibration states.

A commonly used classification of absorption transitions uses the spin states of electrons. Quantum mechanically, the electronic states of atoms and molecules are described by orbitals which define the different states of electrons by two parameters: a geometrical function defining the space and a probability function. The combination of both functions describes the localization of an electron.

Electrons in bonding orbitals are usually paired with antiparallel spin orientation. The total spin S is calculated from the individual electron spins. The multiplicity M is obtained by $M = 2 \times S + 1$. For paired electrons in one orbital these yields:

S = spin (electron 1) + spin (electron 2) = $(+1/2) + (-1/2) = 0$

7.4. LASERS

Laser is an abbreviation for Light Amplification by Stimulated Emission of Radiation. A detailed description of lasers is past the extent of this book.

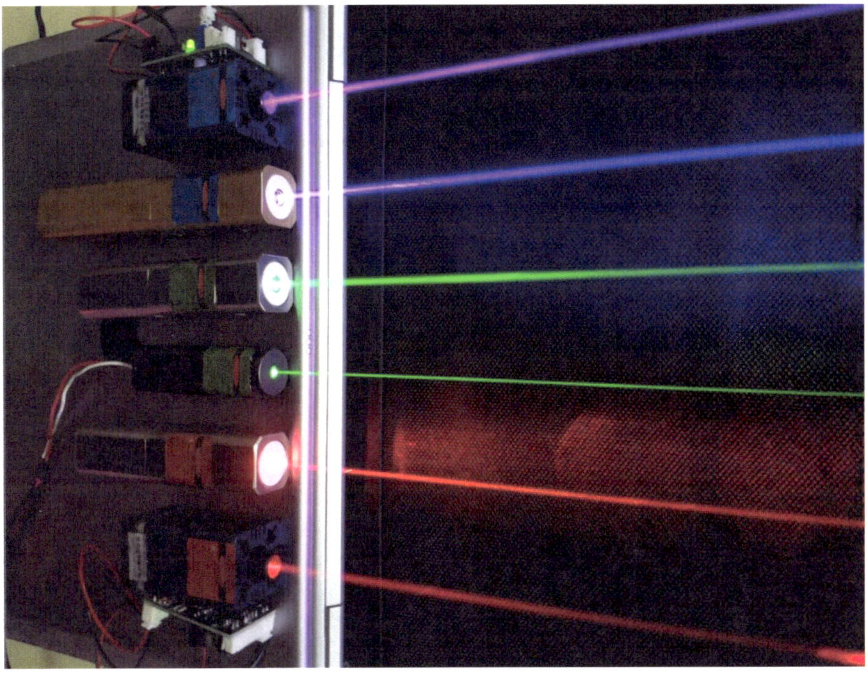

Figure 7.3. Laser beams.

Sources: www.commons.wikimedia.com

This outcome of stimulation is the excitement of an electron to a higher energy level. If, while the electron is in the energized state, one more photon of definitively that energy shows up, then, at that point, rather than the electron being elevated to a significantly more significant level, it can get back to the first ground state. An increase in this cycle will deliver cognizant light with very restricted spectral transmission capacity. To deliver an adequate source of photons, the absorbing material is illuminated by a quickly blazing light of extreme focus (focusing).

Lasers are basic apparatuses in numerous areas of science, including natural chemistry and biophysics. A few current spectroscopic methods use laser light sources, because of their focused energy and precisely characterized spectral properties.

One of the presumably most upsetting applications in the existence sciences, the utilization of lasers in DNA sequencing with fluorescence names, empowered the leap forward in entire genome sequencing.

7.5. ULTRAVIOLET AND VISIBLE LIGHT SPECTROSCOPY

These regions of the electromagnetic spectrum and their associated techniques are probably the most widely used for analytical work and research into biological problems. The electronic transitions in molecules can be classified according to the participating molecular orbitals.

From the four possible transitions ($n \rightarrow \pi^*$, $\pi \rightarrow \pi^*$, $n \rightarrow \sigma^*$, $\sigma \rightarrow \sigma^*$), only two can be elicited with light from the UV/Vis spectrum for some biological molecules: $n \rightarrow \pi$ and $\pi \rightarrow \pi^*$. The $n \rightarrow \sigma^*$ and $\sigma \rightarrow \sigma^*$ transitions are energetically not within the range of UV/Vis spectroscopy and require higher energies.

Molecular (sub-)structures responsible for interaction with electromagnetic radiation are called chromophores. In proteins, there are three types of chromophores relevant for UV/Vis spectroscopy:

Molecular (sub-)structures responsible for interaction with electromagnetic radiation are called chromophores. In proteins, there are three types of chromophores relevant for UV/Vis spectroscopy:

- Peptide bonds (amide bond)
- Certain amino acid side chains (mainly tryptophan and tyrosine)
- Certain prosthetic groups and coenzymes (e.g., coproporphyrins such as in haem).

The presence of several conjugated double bonds in organic molecules results in an extended π-system of electrons which lowers the energy of the π^* orbital through electron delocalization. In many cases, such systems possess $\pi \rightarrow \pi^*$ transitions in the UV/Vis range of the electromagnetic spectrum. Such molecules are very useful tools in colorimetric applications.

7.5.1. Chromophores in Proteins

The electronic transitions of the peptide bond occur in the far UV. The intense peak at 190 nm and the weaker one at 210–220 nm is due to the $\pi \rightarrow \pi^*$ and $n \rightarrow \pi^*$ transitions. Several amino acids (Asp, Glu, Asn, Gln, Arg and His) have weak electronic transitions at around 210 nm.

Usually, these cannot be observed in proteins because they are masked by the more intense peptide bond absorption. The most useful range for proteins is above 230 nm, where there are absorptions from aromatic side chains. While a very weak absorption maximum of phenylalanine occurs at 257 nm, tyrosine and tryptophan dominate the typical protein spectrum with their absorption maxima at 274 nm and 280 nm, respectively.

In praxis, the presence of these two aromatic side chains gives rise to a band at 278 nm. Cystine (Cys_2) possesses a weak absorption maximum of similar strength as phenylalanine at 250 nm. This band can play a role in rare cases of protein optical activity or protein fluorescence.

Proteins that contain prosthetic groups (e.g., haem, flavin, carotenoid) and some metal–protein complexes, may have strong absorption bands in the UV/Vis range. These bands are usually sensitive to the local environment and can be used for physical studies of enzyme action.

Carotenoids, for instance, are a large class of red, yellow and orange plant pigments composed of long carbon chains with many conjugated double bonds. They contain three maxima in the visible region of the electromagnetic spectrum (420 nm, 450 nm, 480 nm).

Porphyrins are the prosthetic groups of hemoglobin, myoglobin, catalase and cytochromes. Electron delocalization extends throughout the cyclic tetrapyrrole ring of porphyrins and gives rise to an intense transition at 400 nm called the Sorbet band. The spectrum of hemoglobin is very sensitive to changes in the iron-bound ligand. These changes can be used for structure-function studies of haem proteins.

Molecules such as FAD (flavin adenine dinucleotide), NADH and NAD^+ are important coenzymes of proteins involved in electron transfer reactions (RedOx reactions). They can be conveniently assayed by using their UV/Vis absorption: 438 nm (FAD), 340 nm (NADH) and 260 nm (NAD^+).

7.6. PRINCIPLES

7.6.1. Quantification of Light Absorption

The chance for a photon to be absorbed by matter is given by an extinction coefficient which itself is dependent on the wavelength ʎ of the photon. If light with the intensity I_0 passes through a sample with appropriate transparency and the path length (thickness) *d*, the intensity *I* drop exponentially drops along with the pathway characteristic absorption parameter for the sample is the extinction coefficient a, yielding the correlation $I= I_0 e^{-ad}$. The ratio $T= I/I_0$ is called transmission.

The Beer-Lambert law is valid for low concentrations only. Higher concentrations might lead to the association of molecules and therefore cause deviations from the ideal behavior. Absorbance and extinction coefficients are additive parameters, which complicates the determination of concentrations in samples with more than one absorbing species.

Note that in dispersive samples or suspensions scattering effects increase the absorbance since the scattered light is not reaching the detector for readout. The absorbance recorded by the spectrophotometer is thus overestimated and needs to be corrected.

7.6.2. Deviations From the Beer-Lambert law

According to the Beer-Lambert law, absorbance is linearly proportional to the concentration of chromophores. This might not be the case anymore in samples with high absorbance. Every spectrophotometer has a certain amount of stray light, which is light received at the detector but not anticipated in the spectral band isolated by the monochromator.

To obtain reasonable signal-to-noise ratios, the intensity of light at the chosen wavelength ($_{I\lambda}$) should be 10 times higher than the intensity of the stray light (I_{stray}). If the stray light gains in intensity, the effects measured at the detector have nothing or little to do with chromophore concentration.

Secondly, molecular events might lead to deviations from the Beer-Lambert law. For instance, chromophores might dimerize at high concentrations and, as a result, might possess different spectroscopic parameters.

7.6.3. Absorption of Light Scattering – Optical Density

In certain applications, for instance, estimation of turbidity of cell colonies (measure of biomass fixation), it isn't the retention yet the dissipating of light that is estimated with a spectrophotometer. Incredibly turbid examples like bacterial colonies don't retain the approaching light.

All things being equal, the light is dissipated and, in this manner, the spectrometer will record an evident absorbance (some of the time likewise called attenuation). For this situation, the noticed boundary is called optical thickness (OD).

Instruments mainly intended to quantify turbid samples are nephelometers or Klett meters; in any case, most biochemical labs utilize the overall UV/Vis spectrometer for measuring of optical densities of cell societies.

7.6.4. Factors Affecting UV/Vi's Absorption

Biochemical samples are usually buffered with aqueous solutions, which has two major advantages. Firstly, proteins and peptides are comfortable in the water as a solvent, which is also the 'native' solvent. Secondly, in the wavelength interval of UV/Vis (700–200 nm), the water spectrum does not show any absorption bands and thus acts as a silent component of the sample.

The absorption spectrum of a chromophore is only partly determined by its chemical structure. The environment also affects the observed spectrum, which mainly can be described by three parameters:

- Protonation/deprotonation (pH, RedOx)
- Solvent polarity (dielectric constant of the solvent)
- Orientation effects.

Vice versa, the immediate environment of chromophores can be probed by assessing their absorption, which makes chromophores ideal reporter molecules for environmental factors. Four effects, two each for wavelength and absorption changes, have to be considered:

- a wavelength shift to higher values is called red shift or bathochromic effect;
- similarly, a shift to lower wavelengths is called blue shift or hypsochromic effect
- an increase in absorption is called hyperchromicity ('more color')
- while a decrease in absorption is called hypo ('less color').

Protonation/deprotonation arises either from changes in pH or oxidation/reduction reactions, which makes chromophores pH- and RedOx- sensitive reporters. As a rule of thumb, λ_{max} and ε increase, i.e., the sample displays a bath- and hyperchromic shift, if a titratable group becomes charged.

Furthermore, solvent polarity affects the difference between the ground and excited states. Generally, when shifting to a less polar environment one observes a bath- and hyperchromic effect. Conversely, a solvent with higher polarity elicits a hypo- and hypochromic effect.

Lastly, orientation effects, such as an increase in the order of nucleic acids from single-stranded to double-stranded DNA, lead to different absorption behavior. A sample of free nucleotides exhibits a higher absorption than a sample with identical amounts of nucleotides but assembled into a single-stranded polynucleotide.

Accordingly, double-stranded polynucleotides exhibit an even smaller absorption than two single-stranded polynucleotides. This phenomenon is called the hypochromic of polynucleotides. The increased exposure (and thus stronger absorption) of the individual nucleotides in the less ordered states provides a simplified explanation for this behavior.

7.7. INSTRUMENTATION

UV/Vi's spectrophotometers are usually dual-beam spectrometers where the first channel contains the sample and the second channel holds the control (buffer) for correction.

On the other hand, one can record the control range first and utilize this as an interior reference for the sample range. The last option approach has become exceptionally well known as numerous spectrometers in the research centers are PC controlled, and correction can be completed utilizing the product by just deducting the control from the example range. The light source is a tungsten fiber bulb for the visible spectrum, and a deuterium bulb for the UV range.

Since the emitted light comprises a wide range of frequencies, a monochromator, comprising either a crystal or an alternating metal lattice of high accuracy called grinding, is put between the light source and the sample. Frequency choice can likewise be accomplished by involving shaded channels as monochromators that retain everything except a specific restricted scope of frequencies. This restricted reach is known as the data transfer capacity of the channel. Channel-based frequency determination is utilized in colorimetry, a technique with moderate exactness, yet the most ideal for basic colorimetric tests where just certain frequencies are of interest.

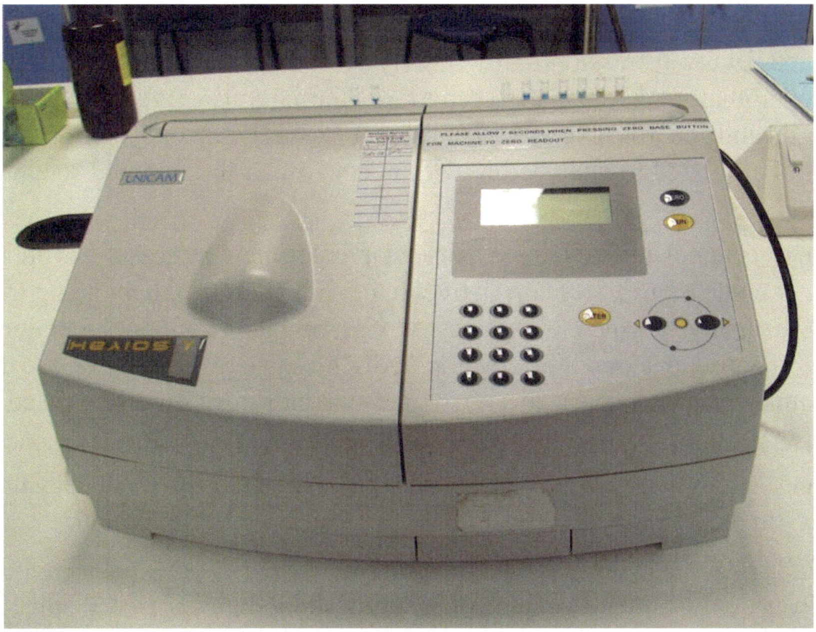

Figure 7.4. Spectro photometer.

Sources: www.commons.wikimedia.com

If frequencies are chosen by crystals or gratings, the strategy is called spectrophotometry. A crystal divides the incident light into its parts by refraction. Refraction happens because the radiation of various frequencies goes along various ways in modes of higher thickness.

To keep up with the standard of speed preservation, the light of more limited frequency (higher speed) should travel a more extended distance (for example blue sky impact). At a grinding, the parting of frequencies is accomplished by diffraction. Diffraction is a reflection peculiarity that occurs at a matrix surface ,

The distance within must be of a similar significant degree as the frequency of the diffracted radiation. By fluctuating the distance hidden therein, various frequencies are chosen. This is accomplished by pivoting the grinding opposite to the optical node.

The goal accomplished by gratings is a lot higher than the one accessible by crystals. These days instruments only contain gratings as monochromators as they can be reproducibly made in great by photo reproduction.

The data transmission a not entirely settled by the channel utilized as monochromator. A channel that seems red to the natural eye is communicating red light and absorbs practically some other (visual) frequency. This channel would be utilized to analyze blue arrangements, as these would retain red light.

The channel utilized for a particular colorimetric measure is in this manner made of a variety correlative to that of the arrangement being tried. Hypothetically, a single frequency is chosen by the monochromator in spectrophotometers, and the developing light is an equal bar. Here, the data transmission is characterized as two times the half-power transfer speed.

The data transmission is an element of the optical cut width. The smaller the cut width the more reproducible are estimated absorbance values. Interestingly, the responsiveness turns out to be less as the cut straight, because less radiation goes through to the locator.

In a dual-beam instrument, the approaching light is parted into two sections by a half mirror. One pillar goes through the sample, the other through a control (blank, reference). This approach forestalls any issues of variety in light force, as both reference and test would be impacted similarly.

The calculated absorbance is the distinction between the two communicated light emissions recorded. Contingent upon the instrument, a subsequent locator estimates the force of the approaching pillar, albeit a few instruments utilize a plan where one indicator estimates the approaching and the sent power on the other hand.

The last option configuration is better according to a logical perspective as it kills possible varieties between the two indicators. At around 350 nm most instruments require a difference in the light source from noticeable to UV light.

This is accomplished by precisely moving mirrors that immediate the proper pillar along with the optical pivot and redirect the other. While examining the time span of 210 nm, this now and again brings about an offset of the range at the switchover point.

Since borosilicate glass and typical plastics retain UV light, such cuvettes must be utilized for applications in the visible range (up to 350 nm). For UV estimations, quartz cuvettes should be utilized. In any case, dispensable plastic cuvettes have been fostered that consider estimations over the whole scope of the UV/Vis range.

7.8. APPLICATIONS

The usual procedure for (colorimetric) assays is to prepare a set of standards and produce a plot of concentration versus absorbance called a calibration curve. This should be linear as long as the Beer-Lambert law applies.

Absorbances of unknowns are then measured and their concentration interpolated from the linear region of the plot. One must never extrapolate beyond the region for which an instrument has been calibrated as this potentially introduces enormous errors.

7.8.1. Qualitative and Quantitative Analysis

Qualitative analysis may be performed in the UV/Vis regions to identify certain classes of compounds both in the pure state and in biological mixtures (e.g., protein-bound). The utilization of UV/Vis spectroscopy for additional logical intentions is fairly restricted, yet feasible for frameworks where suitable standards are known.

Most usually, this kind of spectroscopy is utilized for the measurement of natural samples either straightforwardly or using colorimetric examines. As a rule, proteins can be evaluated straightforwardly utilizing their inherent chromophores, tyrosine and tryptophan. Protein spectra are obtained by filtering from 500 to 210 nm.

The trademark highlights in a protein range are a band at 278/280 nm and one more at 190 nm. The area from 500 to 300 nm gives important data about the presence of any prosthetic gatherings or coenzymes. Protein evaluation by single frequency estimations at 280 and 260 nm just ought to be kept away from, as the presence of bigger totals (impurities or protein totals) brings about significant Rayleigh scattering that should be adjusted.

7.8.2. Difference Spectra

The key benefit of difference spectroscopy is its capacity of detecting a very small absorbance changes in systems with high background absorbance. A difference spectrum is obtained just by subtracting one absorption spectrum from another.

There are two ways by which difference spectra can be obtained: either by subtraction of one absolute absorption spectrum from another, or by placing one sample in the reference cuvette and another in the test cuvette. The latter method needs the application of a dual-beam instrument, while the former method has become very popular because of the most instruments being controlled by computers that eventually allows easy processing as well as handling of data. From a purist's point of view, the direct measurement of the difference spectrum in a dual-beam instrument is the most preferred method, as it reduces the introduction of inconsistencies between samples and therefore, the error of the measurement. The two absolute spectra of ubiquinone and ubiquinol, the oxidised and reduced species of the same molecular skeleton, as well as the difference spectrum.

- difference spectra may have negative absorbance values
- absorption maxima and minima may be displaced and the extinction coefficients are different from those in peaks of absolute spectra
- there are points of zero absorbance, generally accompanied by a slight change of sign of the absorbance values. These points are observed at wavelengths where both species of related molecules depicts identical absorbances (isosbestic points), and which may be applied for checking for the presence of interfering substances.

Common applications for difference UV spectroscopy include the determination of the number of aromatic amino acids exposed to solvent, detection of conformational changes occurring in proteins, detection of aromatic amino acids in active sites of enzymes, and monitoring of reactions involving 'catalytic' chromophores (prosthetic groups, coenzymes).

7.8.3. Derivative Spectroscopy

Another way to resolve small changes in absorption spectra that otherwise would remain invisible is the usage of derivative spectroscopy. Here, the absolute absorption spectrum of a sample is differentiated and the differential plotted against the wavelength. The usefulness of this approach depends on the individual problem. Examples of successful applications include the binding of a monoclonal antibody to its antigen with second-order derivatives and the quantification of tryptophan and tyrosine residues in proteins using fourth-order derivatives.

7.8.4. Solvent Perturbation

As referenced above, aromatic amino acids are the principal chromophores of proteins in the UV locale of the electromagnetic range. Moreover, the UV retention of chromophores relies to a great extent upon the environment.

An adjustment of the extremity of the dissolvable changes the UV range of a protein by bathochromic or hypsochromic impacts without changing its compliance. This peculiarity is called solvent perturbation and can be utilized to measure the outer layer of a protein molecule.

To be available to the solvent, the chromophore must be open on the protein surface. Essentially, solvents like dimethyl-sulfoxide, dioxane, glycerol, mannitol, sucrose and polyethene glycol are utilized for solvent perturbation tests, since they are miscible with water. The technique for solvent perturbation is generally normally utilized for measurement of the number of aromatic residues that are presented to solvents.

7.8.5. Spectrophotometric and Colorimetric Assays

For biochemical measures testing time-or concentrations in reactions , a proper read-out is required that is coupled to the advancement of the response (response coordinate). Hence, the biophysical boundary being checked (read-out) should be coupled to the biochemical boundary being scrutinized.

More often, the monitored parameter is the absorbance of a system at a given wavelength which is monitored throughout the course of the experiment. If possible, one should always try to monitor the changing species directly such as protein absorption, starting product or generated product of a reaction, however in most of the cases this is not possible and a secondary reaction has to be used to generate an appropriate signal for monitoring.

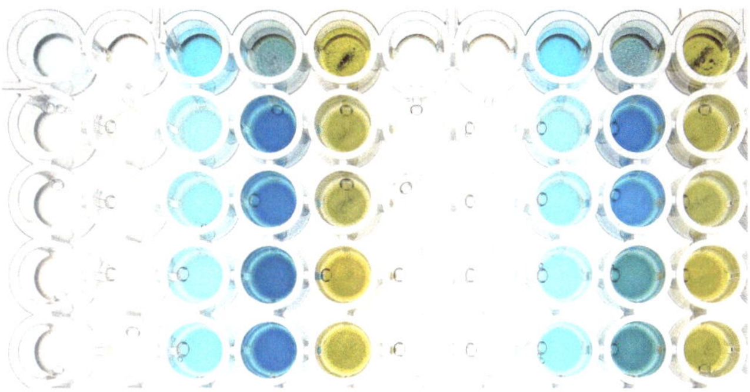

Figure 7.5. A colorimetric assay.

Sources: www.commons.wikimedia.com

A typical use of the last option approach is the measurement of protein focus by Lowry or Bradford measures, where an optional response is utilized to vary the protein. The more extraordinary the variety, the more protein is available. These exams are called colorimetric measures and various regularly utilized.

7.8.6. Fluorescence Spectroscopy

Principles

Fluorescence is an emission phenomenon where an energy transition from a higher to a lower state is accompanied by radiation. Only molecules in their excited forms can emit fluorescence; thus, they have to be brought into a state of higher energy before the emission phenomenon.

Atoms have discrete conditions of the energy. Potential energy levels of particles have been portrayed by various Lennard-Jones curves with overlaid vibrational (and rotational) states. Not all changes are conceivable; permitted advances are characterized by the determination rules of quantum mechanics. A particle in its electronic and vibrational ground state (S0v0) can retain photons matching the energy contrast of its different discrete states.

The necessary photon energy must be higher than that expected to arrive at the vibrational ground condition of the main electronic energized state (S1v0). The excess energy is consumed as vibrational energy ($v > 0$) and immediately disseminated as intensity by a reaction with dissolvable particles.

The atom in this way gets back to the vibrational ground state (S1v0). These unwinding processes are non-transmitting changes starting with one enthusiastic state and then onto the next with lower energy, and are called inner transformation (IC).

From the least level of the principal electronic excited state, the particle gets back to the ground state (S0) either by producing light (fluorescence) or by a non-radiative change. Upon radiative reactions, the particle can wind up in any of the vibrational conditions of the electronic ground state (according to quantum mechanical guidelines).

Assuming the vibrational levels of the ground state cross over with those of the electronic excited state, the particle won't produce fluorescence but instead, return to the ground state by non-radiative inner shifts. This is the most well-known way for excitation energy to be dispersed and is the reason fluorescent atoms are fairly uncommon.

Most atoms are adaptable and, in this way, have extremely high vibrational levels in the ground state. Without a doubt, most fluorescent particles have genuinely unbent aromatic rings or ring frameworks. The fluorescent gathering in a particle is known as a fluorophore.

As radiative energy is lost in fluorescence as compared to the absorption, the fluorescent light is always at a longer wavelength as compared to the exciting light (This is known as Stokes shift). The emitted radiation appears as band spectrum, because there are various closely related wavelength values dependent on the vibrational and rotational energy levels attained.

The fluorescence range of a particle is free of the frequency of the interesting radiation and has a perfect representation relationship with the retention range. The likelihood of the change from the electronic eager to the ground state is corresponding to the power of the radiated light.

An associated phenomenon in this context is phosphorescence which arises from a transition from a triplet state (T_1) to the electronic (singlet) ground state (S_0). The molecule gets into the triplet state from an electronically excited singlet state by a process called intersystem crossing (ISC).

The transition from singlet to triplet is quantum-mechanically not allowed and thus only happens with low probability in certain molecules where the electronic structure is favorable. Such molecules usually contain heavy atoms. The rate constants for phosphorescence are much longer and phosphorescence thus happens with a long delay and persists even when the exciting energy is no longer applied.

The fluorescence properties of a molecule are determined by the properties of the molecule itself (internal factors), as well as the environment of the protein (external factors). The fluorescence intensity emitted by a molecule is dependent on the lifetime of the excited state. The transition from the excited to the ground state can be treated like a decay process of the first order, i.e., the number of molecules in the excited state decreases exponentially with time.

The quantum yield is a dimensionless quantity, and, most importantly, the only absolute measure of fluorescence of a molecule. Measuring the quantum yield is a difficult process and requires comparison with a fluorophore of known quantum yield. In biochemical applications, this measurement is rarely done. Most commonly, the fluorescence emissions of two or more related samples are compared and their relative differences analyzed.

7.9. INSTRUMENTATION

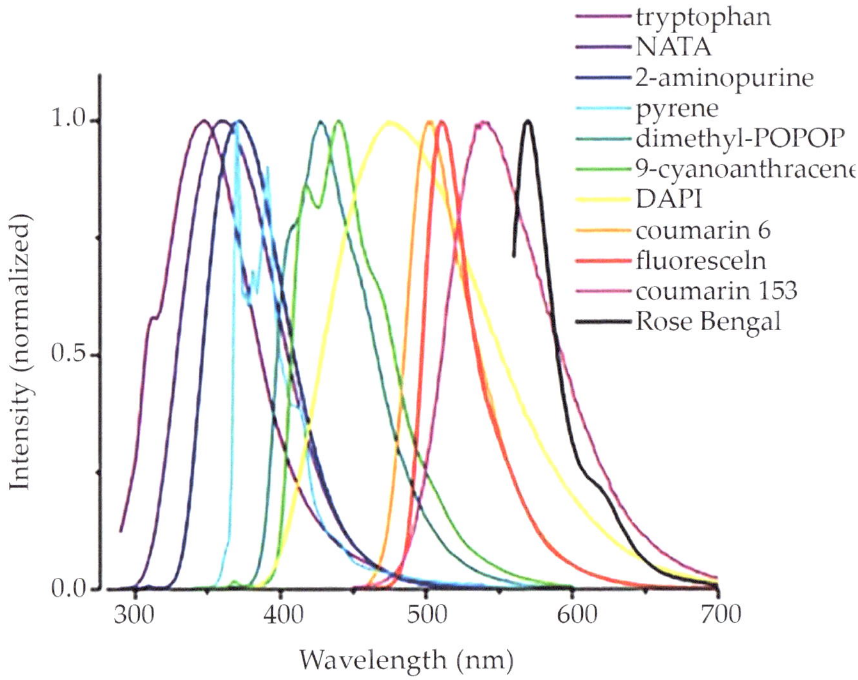

Figure 7.6. Fluorescence Spectroscopy.

Source: www.commons.wikimedia.org

Fluorescence spectroscopy works most precisely at extremely low convergences of radiating fluorophores. UV/Vi's spectroscopy, interestingly, is least precise at such low focuses. One central point adding to the high awareness of fluorescence applications is the otherworldly selectivity.

Because of the Stokes shift, the frequency of the discharged light is not quite the same as that of the interesting light. Another component utilizes the way that fluorescence is discharged every which way. By putting the finder opposite to the excitation pathway, the foundation of the occurrence bar is diminished.

The working of a common spectrofluorometer: Two monochromators are utilized, one for tuning the frequency of the interesting bar and a second one for investigation of the fluorescence discharge. Because the radiated light continuously has lower energy than the intriguing light, the frequency of the excitation monochromator is set at a lower frequency than the discharge monochromator.

The better fluorescence spectrometers in labs have a photon-counting identifier yielding extremely high responsiveness. Temperature control is expected for exact work as the outflow power of a fluorophore is reliant upon the temperature of the arrangement.

Two geometries are possible for the measurement, with the 90° arrangement being most commonly used. Pre- and post-filter effects can take place owing to absorption of light before reaching the fluorophore and the reduction of emitted radiation.

These phenomena are also known as inner filter effects. They are more evident in solutions with high concentrations. As a rough guide, the absorption of a solution to be used for fluorescence experiments should be less than 0.05. The use of microcuvettes containing less material can also be useful. On the other hand, the front-face illumination geometry can be used which obviates the inner filter effect.

In addition, while the 90° geometry needs cuvettes with two neighbouring faces being clear (in general, fluorescence cuvettes have four clear faces), the front-face illumination technique requires only one clear face, as excitation and emission occur at the same face. Though, front-face illumination is less sensitive than the 90° illumination.

7.10. APPLICATIONS

There are many and highly varied applications for fluorescence even though relatively few compounds exhibit the phenomenon. The effects of pH, solvent composition and the polarization of fluorescence may all contribute to structural elucidation. Measurement of fluorescence lifetimes can be used to assess rotation correlation coefficients and thus particle sizes.

Non-fluorescent compounds are often labeled with fluorescent probes to enable the monitoring of molecular events. This is termed extrinsic fluorescence as distinct from intrinsic fluorescence where the native compound exhibits the property. Some fluorescent dyes are sensitive to the presence of metal ions and can thus be used to track changes of these ions' in *in vitro* samples, as well as whole cells.

Since fluorescence spectrometers have two monochromators, one for tuning the excitation wavelength and one for analyzing the emission wavelength of the fluorophore, one can measure two types of spectra: excitation and emission spectra.

For fluorescence excitation spectrum measurement, one sets the emission monochromator at a fixed wavelength (λ_{em}) and scans a range of excitation wavelengths that are then recorded as ordinate (x-coordinate) of the excitation spectrum; the fluorescence emission at λ_{em} is the abscissa.

Measurement of emission spectra is achieved by setting a fixed excitation wavelength (lex) and scanning a wavelength range with the emission monochromator. To yield a spectrum, the emission wavelength λ_{em} is recorded as ordinate and the emission intensity at λ_{em} is as abscissa.

7.10.1. Intrinsic Protein Fluorescence

Proteins have three inherent fluorophores: tryptophan, tyrosine and phenylalanine, albeit the last option has an extremely low quantum yield and its commitment to protein fluorescence outflow is in this manner unimportant. Of the excess two deposits, tyrosine has the lower quantum yield and its fluorescence discharge is as a rule extinguished when it becomes ionized, or is situated close to an amino or carboxyl gathering, or a tryptophan buildup.

Characteristic protein fluorescence is not entirely settled by tryptophan fluorescence which can be specifically excited at 295-305 nm. Excitation at 280 nm energizes tyrosine and tryptophan fluorescence and the subsequent spectra could hence contain contributions from both types of residues.

The primary application for characteristic protein fluorescence focuses on conformational measurements. We have previously referenced that the fluorescence properties of a fluorophore rely fundamentally upon environmental variables, including solvents, pH, possible quenchers, neighbouring groups, etc.and so forth.

Several empirical rules can be applied to interpret protein fluorescence spectra:

- As a fluorophore moves into an environment with less polarity, its emission spectrum exhibits a hypsochromic shift (λ_{max} moves to shorter wavelengths) and the intensity at λ_{max} increases.
- Fluorophores in a polar environment show a decrease in quantum yield with increasing temperature. In a non-polar environment, there is little change.

- Tryptophan fluorescence is quenched by neighboring protonated acidic groups.

When interpreting effects observed in fluorescence experiments, one has to consider carefully all possible molecular events. For example, a compound added to a protein solution can cause quenching of tryptophan fluorescence.

This could come about by binding of the compound at a site close to the tryptophan (i.e., the residue is surface exposed to a certain degree), or due to a conformational change induced by the compound. The comparison of protein fluorescence excitation and emission spectra can yield insights into the location of fluorophores.

The close spatial arrangement of fluorophores within a protein can led to the quenching of fluorescence emission; this might be seen by the lower intensity of the emission spectrum when compared to the excitation spectrum.

7.10.2. Extrinsic Fluorescence

Regularly, atoms of interest for biochemical investigations are non-fluorescent. In large numbers of these cases, an outside fluorophore can be brought into the framework by synthetic coupling or non-covalent restricting.

Three standards should be met by fluorophores in this unique circumstance. First and foremost, it should not influence the structural properties of the framework being scrutinized. Also, its fluorescence discharge should be sensitive to environmental circumstances to allow the checking of the molecular events. Furthermore, in conclusion, the fluorophore should be firmly bound at an extraordinary area.

A typical chromophore for proteins is 1-aniline-8-naphthalene sulphonate (ANS), which emanates just feeble fluorescence in polar surroundings, for example in the fluid arrangement. Be that as it may, in a non-polar surrounding, for example at the point when bound to hydrophobic patches on proteins, its fluorescence discharge is essentially expanded and the range shows a hypsochromic shift; λmax shifts from 475 nm to 450 nm.

ANS is thus a valuable tool used for assessing the degree of non-polarity. It can also be utilizied in competition assays so as to monitor binding of ligands and prosthetic groups.

Metal-chelating compounds with fluorescent properties are valuable devices for an assortment of tests, and metal homeostasis for cells. Generally tests for calcium are the chelators Fura-2, Indo-1 and Quin-1. Since the science of such mixtures depends on metal chelation, cross-reactivity of the tests with other metal particles is conceivable.

The characteristic fluorescence of nucleic acids is exceptionally feeble and the expected excitation frequencies are excessively far in the UV district to be helpful for pragmatic applications. Various outward fluorescent tests precipitously tie to DNA and show improved discharge.

While in prior days ethidium bromide was one of the most broadly involved dyes for this application, it has these days been supplanted by SYBR Green, as the latter option presents fewer hazards for health and the environment and has no teratogenic properties like ethidium bromide.

These tests bind DNA by intercalation of the planar aromatic ring frameworks between the base sets of twofold helical DNA. Their fluorescence emanation in water is exceptionally powerless and increments around 30-overlap after restricting to DNA.

7.10.3. Fluorescence Resonance Energy Transfer (FRET)

Fluorescence resonance energy transfer (FRET) was first described by Forster in 1948. The process can be explained in terms of quantum mechanics by a non-radiative energy transfer from a donor to an acceptor chromophore.

The requirements for this process are a reasonable overlap of emission and excitation spectra of donor and acceptor chromophores, close spatial vicinity of both chromophores (10–100 A°), and an almost parallel arrangement of their transition dipoles. Of great practical importance is the correlation g that the FRET effect is inversely proportional to the distance between donor and acceptor chromophores, R_0.

The FRET impact is especially appropriate for organic applications since distances of 10-100 A° are in the request for the components of natural macromolecules. Moreover, the connection between FRET and the distance considers the estimation of sub-atomic distances and makes this application a sort of ‘spectroscopic ruler’. If a cycle shows changes in sub-atomic distances, FRET can likewise be utilized to screen the atomic components.

Much of the time, various chromophores are utilized as donors and acceptors, introducing two prospects to record FRET: either as contributor excited fluorescence outflow of the acceptor or as fluorescence extinguishing of the giver by the acceptor.

Be that as it may, a similar chromophore might be utilized as a contributor and acceptor all the while; for this situation, the depolarization of fluorescence is the noticed boundary. Since non-FRET excited fluorescence emanation by the acceptor can bring about unwanted background fluorescence, a typical methodology is to use the non-fluorescent acceptor chromophores.

FRET-based tests might be utilized to explain the impacts of new substrates for various compounds or putative agonists speedily and quantitatively. Moreover, FRET discovery may be utilized in high-throughput screenings, which makes it extremely alluring for drug advancement.

7.10.4. Bioluminescence Resonance Energy Transfer (BRET)

Bioluminescence resonance energy transfer (BRET) uses the FRET effect with native fluorescent or luminescent proteins as chromophores. The phenomenon is observed naturally, for example with the sea pansy *Renilla reniformis*. It contains the enzyme luciferase, which oxidizes luciferin (coelenterazine) by simultaneously emitting light at λ_{exc} = 480 nm. This light directly excites green fluorescent protein (GFP), which, in turn, emits fluorescence at λ_{em} = 509 nm.

Fluorescence naming of proteins by different proteins presents a helpful way to deal with concentrating different cycles in vivo. Marking should be possible at the hereditary level by creating combination proteins. Checking of protein articulation by GFP is a laid-out method and further advancement of 'living varieties' will prompt promising new devices.

While nucleic acids have been the principal players in the genomic period, the postgenomic/proteomic time focuses on quality items, the proteins. New proteins are being found and described; others are as of now utilized inside biotechnological processes.

Specifically for the characterization and assessment of proteins and receptors, response systems can be planned to such an extent that the response of interest is recognizable quantitatively utilizing FRET contributor and acceptor matches.

For example, identification strategies for protease action can be created given BRET applications. A protease substrate is bound to a GFP variation on the N-terminal side and dsRED on the C-terminal side. The last protein is a red fluorescing FRET acceptor and the GFP variation goes about as a FRET contributor.

When the substrate is targeted by a protease, the FRET impact is nullified. This is utilized to screen protease action directly. With a blend of FRET examination and two-photon excitation spectroscopy completing a motor analysis is likewise conceivable.

A comparable thought is utilized to name human insulin receptors to survey its movement quantitatively. The insulin receptor is a glycoprotein with two on and two b subunits, which are connected by thioether spans. The limiting of insulin incites a conformational change and causes a nearby spatial plan of both b subunits. This, thus, initiates tyrosine kinase movement of the receptor.

In some circumstances, for example, diabetes, the tyrosine kinase action is not the same as in regular circumstances. It is of extraordinary interest to find builds that stimulate a similar action as insulin. By binding the b subunit of human insulin receptor to *Renilla reniformis* luciferase and yellow fluorescent protein (YFP) a FRET giver acceptor pair is gotten, which reports the ligand-prompted conformational change and goes before the sign transduction step. This system can distinguish the impacts of insulin and insulin-stimulating ligands to evaluate protein behavior.

7.10.5. Fluorescence Recovery after Photobleaching (FRAP)

On the off chance that a fluorophore is presented to extreme incident radiation, it could be irreversibly harmed and lose its capacity to discharge fluorescence. Purposeful bleaching of a negligible portion of fluorescently marked particles in a layer can be utilized to screen the movement of named particles in certain (two-layered) compartments.

Furthermore, the time-dependent monitoring permits the determination of the diffusion coefficient. A well-established application is the application of phospholipids labelled with NBD (e.g. NBD-phosphatidylethanolamine, which are combined into a biological or artificial membrane. The specimen is subjected to a pulse of high-intensity light (photo bleaching), which causes a sharp drop of fluorescence in the observation area.

Re-emergence of fluorescence emission in this area is also observed as unbleached molecules diffused into the observation area. From the time-dependent increase of fluorescence emission, the rate of diffusion of the phospholipid molecules can be calculated. In the same manner, membrane proteins such as receptors or even proteins in a cell can be conjugated to fluorescence labels and their diffusion coefficients can be determined.

7.10.6. Fluorescence Polarization

A light source normally comprises an assortment of randomly occuring producers, and the transmitted light is an assortment of waves with all potential directions of the E vectors (non-captivated light). Straightly energized light is gotten by going light through a polarizer that communicates light with just a solitary plane of polarization; for example, it passes just those parts of the E vector that are lined up with the center of the polarizer.

The force of communicated light relies upon the direction of the polarizer. The greatest transmission is accomplished when the plane of polarization is lined up with the axis of the polarizer; the transmission is zero when the direction is opposite.

Tentatively, it can be accomplished in a fluorescence spectrometer by putting a polarizer in the excitation way to stimulate the sample with polarised light. A second polarizer is set between the example and the locator with its pivot either equal or opposite to the hub of the excitation polarizer. The produced light is either somewhat enraptured or completely unpolarized. This deficiency of polarization is called fluorescence depolarization.

Accordingly, if the chromophores are haphazardly arranged in arrangement, the polarization P is under 0.5. It is subsequently clear that any interaction that prompts a deviation from an irregular direction will lead to a difference in polarization. This is positively the situation when a chromophore turns out to be more static.

Moreover, one necessity to consider the Brownian movement. If the chromophore is a little particle in arrangement, it will turn quickly. Any adjustment of this movement because of temperature changes, changes in the consistency of the dissolvable, or restricted to a bigger particle, will subsequently bring about a difference in polarization.

7.10.7. Fluorescence Cross-Correlation Spectroscopy

With fluorescence cross-correlation spectroscopy the temporal fluorescence fluctuations between two differently labeled molecules can be measured as they diffuse through a small sample volume. Cross-correlation analysis of the fluorescence signals from separate detection channels extracts information on the dynamics of the dual labeled molecules.

Fluorescence cross-correlation spectroscopy has thus become an essential tool for the characterization of diffusion coefficients, binding constants, kinetic rates of binding and determining molecular interactions in solutions and cells.

7.10.8. Fluorescence Microscopy, High-Throughput Assays

Fluorescence microscopy for checking is a significant instrument for some organic and biochemical applications. We have previously seen the utilization of fluorescence in DNA sequencing; the strategy is indistinguishably connected to the outcome of activities, for example, genome sequencing. Fluorescence procedures are likewise crucial techniques for cell applications with fluorescence microscopy.

Proteins (or natural macromolecules) of interest can be labeled with a fluorescent tag, for example, the green fluorescent protein (GFP) from the jellyfish *Aequorea victoria* or the red fluorescent protein from *Discosoma striata*, for spatial tracking of the labeled protein.

On the other hand, the utilization of GFP spectral variations like cyan fluorescent protein (CFP) as a fluorescence donor and yellow fluorescent protein (YFP) as an acceptor permits examination of structural studies by utilizing the FRET. Samples with cells expressing the tagged proteins are illuminated with the light of the excitation frequency and afterward observed through a channel that omits the absorbing light and just sends the fluorescence signal.

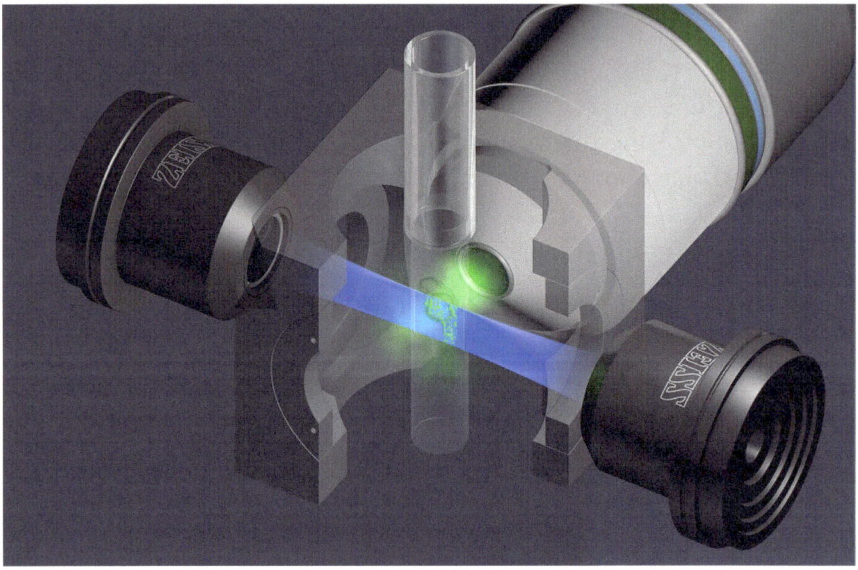

Figure 7.7. Fluorescence chromatographer-its working.

Source: Image by flickr.com

The recorded fluorescence signal can be overlaid with a visual picture computationally, and the composite picture then, at that point, takes into consideration the localization of the tagged species. Assuming different fluorescence marks with specific emission frequencies are utilized all the while, even co-localization studies can be performed.

7.10.9. Time-Resolved Fluorescence Spectroscopy

By sampling the photon emission for a large number of excitations, the probability distribution can be constructed. The time-dependent decay of an individual fluorophore species follows an exponential distribution, and the time constant is thus termed the lifetime of this fluorophore.

Curve fitting of fluorescence decays enables the identification of the number of species of fluorophores (within certain limits), and the calculation of the lifetimes of these species. In this context, different species can be different fluorophores or distinct conformations of the same fluorophore.

7.11. CONCLUSION

The chapter discussed the various spectroscopy techniques that have been used in biochemistry. It also discussed the several properties of electromagnetic radiation. This chapter also discussed the concept of interaction with matter, and the concept of laser. Towards the end of the chapter, it explains UV and visible light spectroscopy and their principles.

REFERENCES

1. Bruch, M., 1996. NMR Spectroscopy Techniques. [online] Available at: <https://www.taylorfrancis.com/books/edit/10.1201/9781482273311/nmr-spectroscopy-techniques-martha-bruch> [Accessed 4 July 2022].
2. Joseph, A., 2020. Neutron spectroscopy techniques. *Micro and Nanostructured Composite Materials for Neutron Shielding Applications*, [online] pp.355-378. Available at: <https://www.sciencedirect.com/science/article/pii/B9780128194591000131?via%3Dihub> [Accessed 4 July 2022].
3. Krivanek, O., 1989. Spectroscopy techniques. *Ultramicroscopy*, [online] 28(1-4), pp.213-214. Available at: <https://linkinghub.elsevier.com/retrieve/pii/0304399189902982> [Accessed 4 July 2022].
4. Morrison, K., 2019. Spectroscopy techniques. *Characterisation Methods in Solid State and Materials Science*, [online] Available at: <https://iopscience.iop.org/book/978-0-7503-1383-4/chapter/bk978-0-7503-1383-4ch5> [Accessed 4 July 2022].
5. White, S., 2011. Principles and techniques of biochemistry and molecular biology, seventh edition. *Biochemistry and Molecular Biology Education*, [online] 39(3), pp.244-245. Available at: <https://iubmb.onlinelibrary.wiley.com/doi/10.1002/bmb.20499> [Accessed 4 July 2022].

CHAPTER 8

BIOCHEMISTRY OF LIPIDS

CONTENTS

Including fats, oils, hormones, and certain components of membranes that are grouped together, lipids, are a diverse group of organic compounds because they do not interact appreciably with water. As fat in adipose cells, one type of lipid, the triglycerides provides thermal insulation and serves as the energy-storage depot for organisms.

Between cells, tissues, and organs, some lipids such as steroid hormones serve as chemical messengers and others between biochemical systems within a single cell communicate signals. The membranes of cells and organelles (structures within cells) are microscopically thin structures formed from two layers of phospholipid molecules.

8.1. INTRODUCTION

To separate individual cells from their environments and to compartmentalize the cell interior into structures, membranes function to carry out special functions. Membranes, and the lipids that form them, must have been essential to the origin of life itself so important is this compartmentalizing function.

Lipids are ancient and ubiquitous molecules. Differences are found in the lipid chemistry of the predominant building blocks among the three domains of life on our planet, (e.g., l-glycerol vs. d-glycerol, ester vs. ether linkages, among others) but even many viruses possess lipid envelopes until they are shed inside the host cell, between life forms.

One of the defining characteristics of an organism is the evolution of an outer membrane composed of a complex mixture of lipids, proteins, and carbohydrates. From two basic biosynthetic pathways, lipids are generated. Derived from malonyl-CoA and acetyl-CoA esters and a carbanion intermediate, the first involves the condensation of acyl carrier protein intermediates.

This pathway leads to diverse classes of lipids that contain fatty acyl chains, including fatty acids, phospholipids, and glycolipids. A similar pathway in plants is provided by the polyketide biosynthetic pathway. The condensation of branched-chain five-carbon pyrophosphate intermediates and a carbocation intermediate are involved in the second biosynthetic pathway.

In the Archaea domain as well as several species in the Bacteria and Eukarya domains, this latter pathway is the source of all lipid species identified, such as sterols, archaeal glycolipids and prenols, sphingolipids and glycerophospholipids.

By some investigators to be a more enlightened definition of what molecules are lipids, appreciation of these biosynthetic sources has been suggested, in an organic solvent as opposed to classic definitions of solubility (reviewed in Brown, H.A.; Murphy R.C. Nat. Chem. Biol.2009, 5, 602–606). To become involved in the multitudes of biological processes used by living organisms, it is clear that lipids evolved from these biosynthetic pathways.

When one considers chirality, precise locations of double bonds, attachments of various head groups, carbohydrates, and amino acids, and other potential chemical diversity, the numbers are in the thousands or beyond because the comprehensive counting of the total number of lipid molecular species in nature has yet to be fully tallied.

In both structure and biological function lipids as a class of molecules display a wide diversity. The primary roles of lipids in cellular function are the formation of the permeability barrier of cells and subcellular organelles in the form of a lipid bilayer.

In almost all membranes, glycerol-based phospholipid is the major lipid type defining this bilayer, although other lipids are important components and vary in their amounts and presence across the spectrum of organisms. In a few bacterial membranes and in all eukaryotic cytoplasmic membranes, sterols are present.

In the membranes of all eukaryotes, the ceramide-based sphingolipids are also present. While Gram-negative bacteria utilize a glucosamine-based phospholipid (Lipid A) as a major structural component of the outer membrane; neutral glycerol-based glycolipids are major membrane-forming components in many Gram-positive bacteria and in the membranes of plants.

Additional diversity results in the variety of the hydrophobic domains of lipids . With varying numbers and positions of double bonds, in eukaryotes and eubacteria, these domains are usually long chain fatty acids or alkyl alcohols. The phospholipids have long chain reduced polyisoprene moieties in the case of archaebacteria, rather than fatty acids, in ether linkage to glycerol.

The number of individual phospholipid species ranges in the hundreds if one considers a simple organism such as *Escherichia coli* with three major phospholipids and several different fatty acids along with many minor precursors and modified products. In both the phospholipids and fatty acids, the number of individual species is in the thousands with greater diversity in more complex eukaryotic organisms..

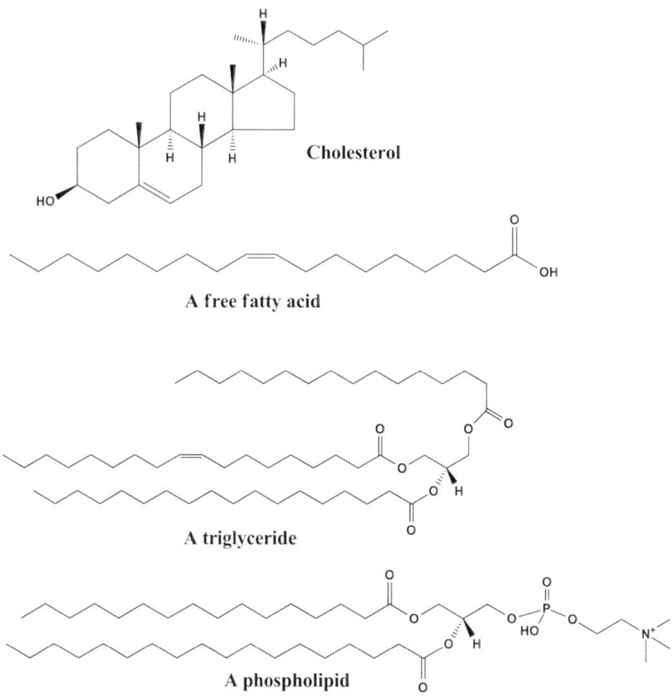

Figure 8.1. Image showing 4 common lipids.

Source: Image by Wikimedia commons

To form a stable bilayer structure if one or two phospholipids are sufficient, why is there the above diversity in lipid structures present in biological membranes? Only with a broad spectrum of lipid mixtures the adaptability and flexibility in membrane structure necessitated by the environment is possible.

For a wide spectrum of proteins involved in many cellular processes, the membrane is also the supporting matrix. Probably half of the remaining proteins function at or near a membrane surface and approximately 20-35% of all proteins are integral membrane proteins.

Therefore, the role of lipids dynamic with respect to cell function rather than simply defining a static barrier made by the physical and chemical properties of the membrane directly affects most cellular processes making.

8.2. DIVERSITY IN LIPID STRUCTURE

Those biological molecules readily soluble in organic solvents such as chloroform, ether, or toluene are known as lipids. However, lipids with large hydrophilic domains such as lipopolysaccharide are not soluble in these solvents and some very hydrophobic proteins such as the F0 subunits of ATP synthase are soluble in chloroform. Those lipids that contribute significantly to membrane structure or have a role in determining protein structure or function are considered only.

8.2.1. Glycerol-Based Lipids

The primary building blocks of most membranes are glycerol phosphate-containing lipids generally referred to as phospholipids. Sn-3-glycerol esterified at the 1- and 2-position with long chain fatty acids is the diacylglycerol backbone in eubacteria and eukaryotes. In archaebacteria, sn-l-glycerol forms the lipid backbone, and the hydrophobic domain is composed of phytanyl (a saturated iso-prenyl) groups in ether linkage at the 2- and 3-position (an archaeol).

By two bi-phytanyl groups (di-bi-phytanyl-di-glycero-phosphate-tetraether) in addition two sn-l-glycerol groups are found connected in ether linkage to form a covalently linked bilayer. In the plasmalogens of eukaryotes, some eubacteria (mainly hyperthermophiles) have diallyl (long-chain alcohols in ether linkage) glycerophosphate lipids and similar ether linkages are found.

The headgroups of the phospholipids extend the diversity of lipids by phosphatidylcholine (PC), phosphatidic acid (PA, with OH), phosphatidylglycerol (PG), phosphatidylserine (PS), phosphatidylinositol (PI), and cardiolipin (CL). Archaebacteria analogs exist with headgroups of glycerol and glycerol methyl phosphate as well as all of the above except PC and CL.

Mono- and disaccharides (glucose or galactose) are directly linked to sn-l-archaeol archaebacteria and also have neutral glycolipid derivatives. Plants (mainly in the thylakoid membrane) and many Gram-positive bacteria also have high levels of neutral glycolipids with mono or disaccharides linked to the 3-carbon of *sn*-3-diacylglycerol.

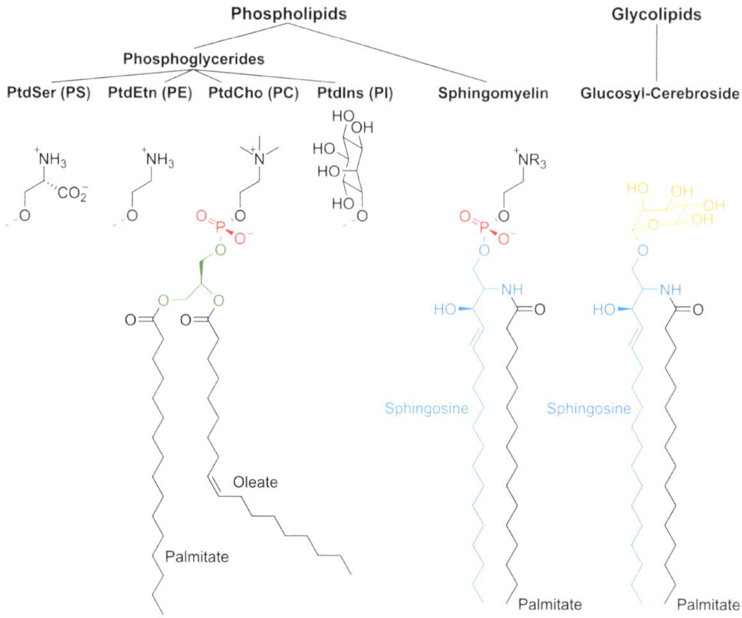

Figure 8.2. Scheme of the general structures of membrane phospholipids and glycolipids.

Source: Image by Wikimedia commons

Therefore, in a single organism is significant the diversity of glycerol-based lipids, but the diversity throughout nature is enormous. With only limited information on the lipids from archaebacteria, most of the information on the chemical and physical properties of lipids comes from studies on the major phospholipid classes of eubacteria and eukaryotes.

In eubacteria and eukaryotes, the biosynthetic pathways and the genetics of lipid metabolism have also been extensively studied. With respect to the environment of archaebacteria clearly, the archaeol lipids confer some advantage. Many of these organisms exist in harsh environments that call for more chemically stable lipid bilayers, which is afforded by the above lipids.

8.2.2. Di-Glucose-Amine Phosphate-Based Lipids

A lipid made up of a headgroup derived from glucosamine phosphate is contained in the outer membrane of Gram-negative bacteria. The core lipid in *E. coli* is a phospholipid containing two glucosamine groups in β(1-6) linkage that have at positions 2, 3, 2' and 3' with *R*-3-hydroxymyristic acid (C 14) and at positions 1 and 4' with phosphates.

Further modification at position 6' by an inner core, an outer core, and the O-antigen with a KDO disaccharide (two 3-dcoxy-D-manno-octulosonic acids in α (1-3) linkage) results in KDO2-Lipid A.

As lipopolysaccharide or LPS, the complete structure is either with or without O-antigen. The inner monolayer of the outer membrane is made up of glycerophosphate-based lipids, the core Lipid A forms the outer monolayer of the outer membrane bilayer of Gram-negative bacteria.

Only the KDO_2-Lipid A structure is essential for the viability of laboratory strains. The whole lipopolysaccharide structure defines the outer surface of Gram-negative bacteria.

However, for the survival of Gram-negative bacteria in their natural environment the remainder of the lipopolysaccharide structure is important. In response to environment including host fluids, temperature, ionic properties, and antimicrobial agents, this structure is modified post-assembly.

In all component parts of the LPS structure, both enteric and non-enteric Gram-negative bacteria show a great diversity. Because it is the primary antigen responsible for toxic shock syndrome caused by Gram-negative bacterial infection studies of Lipid A biosynthesis is of clinical importance.

8.3. PROPERTIES OF LIPIDS IN SOLUTION

A lipid bilayer composed of a hydrophobic core excluded from water and an ionic surface that interacts with water and defines the hydrophobic-hydrophilic interface is a biological membrane defined by the matrix. Much of our understanding of the physical properties of lipids in solution and the driving force for the formation of lipid bilayers comes from the concept of the 'hydrophobic effect' as developed by Charles Tanford.

The 'fluid mosaic' model for membrane structure further popularized these concepts. Since It envisioned membrane proteins as undefined globular structures freely moving in a homogeneous sea of lipids.

8.3.1. Why Do Polar Lipids Self-Associate?

Polar lipids are amphipathic in nature containing both hydrophilic domains that readily interact with water and hydrophobic domains, which do not interact with water. The basic premise of the hydrophobic effect is that the hydrocarbon domains of polar lipids disrupt the stable hydrogen-bonded structure of water and when such domains self-associate to minimize the total surface area in contact with water, therefore, are at an energy minimum.

The polar domains of lipids interact either through ionic interaction with water or hydrogen bonding and are energetically stable in an aqueous environment. By its concentration and the law of opposing forces, the structural organization that a polar lipid assumes in water is determined i.e., steric and ionic repulsive forces of the polar domains versus hydrophobic forces driving self-association in opposing self-association.

Amphipathic molecules exist as monomers in solution, at low concentrations. Its stability in solution as monomers decrease, as the concentration of the molecule increases, until the favorable interaction of the polar domain with water is outweighed by the unfavorable interaction of the hydrophobic domain with water.

At this point, a further increase in concentration results in the formation of increasing amounts of self-associated monomers in equilibrium with a constant amount of free monomers. The remaining constant free monomer concentration and this point of self-association is the critical micelle concentration. Due to the increased hydrophobic effect the larger the hydrophobic domain, the lower the critical micelle concentration.

However, either because of the size of neutral domains or charge repulsion for ionic domains, the larger the polar domain, the higher the critical micelle concentration due to the unfavorable steric hindrance in bringing these domains into proximity.

The critical micelle concentration of amphipathic molecules with a net charge is influenced by the ionic strength of the medium due to dampening of the charge repulsion effect. Therefore, when the NaC1 concentration is raised from 0 to 0.5 M, the critical micelle concentration of the detergent sodium dodecyl sulfate is reduced ten-fold.

Three supramolecular structural organizations of polar lipids and detergents in solution are defined by these physical properties and the shape of amphipathic molecules. Detergents, lyso-phospholipids (containing only one alkyl chain), and phospholipids with short alkyl chains (eight or fewer carbons) have an inverted cone-shape (large head group relative to a small hydrophobic domain) and self-associate above the critical micelle concentration with a small radius of curvature with a hydrophobic core excluding water to form micellar structures.

The micelle surface is a very rough surface with many of the hydrophobic domains exposed to water rather than being a smooth spherical or elliptical structure with the hydrophobic domains completely sequestered inside a shell of polar residues that interact with water.

By balancing the attractivc forcc of the hydrophobic effect and the repulsive force of close headgroup association the overall structure reflects the packing of amphipathic molecules at an energy minimum. Most detergents have micromolar to millimolar critical micelle concentrations. With critical micelle concentrations in the micromolar range, lyso-phospholipids also form micelles.

However, due to the hydrophobic driving force contributed by two alkyl chains, phospholipids with chain lengths of 14 and above self - associate at a concentration around 10^{-10} M. Phospholipids with long alkyl chains do not form micelles but organize into bilayer structures, with the maximum exclusion of water from the hydrophobic domain, which allows tight packing of adjacent side chains.

Phospholipids are not found free as monomers in solution but are organized into either membrane bilayers or protein complexes in living cells. They spontaneously form large multilamellar bilayer sheets separated by water when long chain phospholipids are first dried to a solid from organic solvent and then hydrated. By closing into sealed vesicles, sonication disperses these sheets into smaller uni-lamellar bilayer structures that satisfy the hydrophobic nature of the ends of the bilayer (also termed liposomes) defined by an aqueous core and a continuous single bilayer much like the membrane surrounding cells.

Through a small orifice or by dilution of a detergent-lipid mixture below the critical micelle concentration of the detergent, liposomes can also be made by physical extrusion of lamellar structures. PC forms lipid bilayers that are cylindrical-shaped lipids (hydrophobic domains and head groups of similar diameters).

PE (unsaturated fatty acids) favors an inverted micellar structure where the headgroups sequester an internal aqueous core and the hydrophobic domains are oriented outward and self-associate in non-bilayer structures are cone-shaped lipids (small head groups relative to a large hydrophobic domain).

These are denoted as the cubic phases (a more complex organization like the H_{II} phase) and hexagonal II (H_{II}). Lipid polymorphism is referred to as the ability of lipids to form multiple structural associations. Depending on solvent conditions, alkyl chain composition, and temperature, lipids such as PE, PA, CL, and monosaccharide derivatives of diacylglycerol can exist in either bilayer or the H_{II} phase.

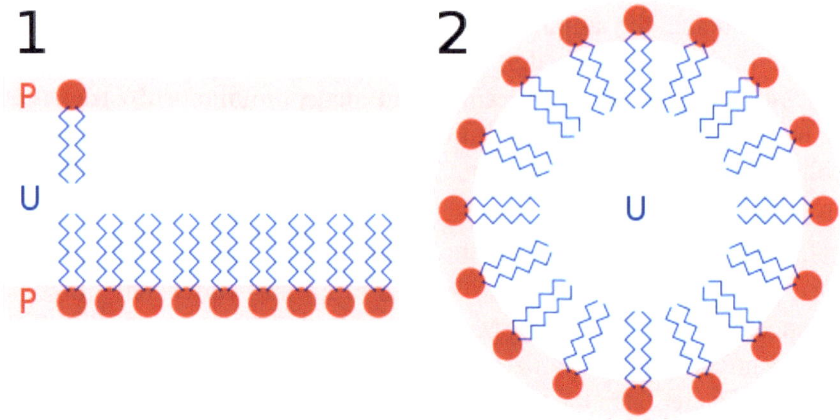

Figure 8.3. Lipid bilayer and micelle.

Source: Image by Wikimedia commons

When mixed with the bilayer-forming lipids change the physical properties of the bilayer and introduce stress or strain in the bilayer structure. Both cone-shaped and inverted cone-shaped lipids are considered as non-bilayer forming lipids. When they are spread as a monolayer at an aqueous-air interface, bilayer-forming lipids have no tendency to bend away from or toward the aqueous phase due to their cylindrical symmetry. The hydrophobic domain orients toward the air, in such a system.

Monolayers of asymmetric inverted cone-shaped lipids (micelle-forming) tend to bend away from the aqueous phase (positive radius of curvature) while monolayers of the asymmetric cone-shaped lipids (H_{II} -forming) tend to bend toward from the aqueous interface (negative radius of curvature).

8.3.2. Special Properties of Cardiolipin

Depending on the absence or presence of divalent cations, CL has the unique property of being both a bilayer and non-bilayer lipid. CL is found almost exclusively in eukaryotic mitochondria and in bacteria that utilize oxidative phosphorylation for proton pumping across the membrane.

The ionization constants of its two phosphate diesters is a property of CL that has gone largely unrecognized. In the range of 2-4, pK_2 of CL is >8.5, rather than displaying two pK values indicating that CL is protonated at physiological pH. In transfer processes, this property may make CL a proton sink or a conduit for protons.

Lack of CL results in a reduction in cell growth dependent on oxidative processes although PG appears to substitute for CL in many processes in both bacteria and yeast. Therefore, CL appears to be required for optimal cell function, but it is not essential.

8.4. ENGINEERING OF MEMBRANE LIPID COMPOSITION

How can the role of a given lipid be defined at the molecular level, given the diversity in both lipid structure and function? Alone, lipids have neither inherent catalytic activity nor obvious functions unlike proteins (except for their physical organization). But by their effect on catalytic processes or biological functions studied *in vitro*, many functions of lipids have been uncovered serendipitously. Such studies are highly prone to artifacts, although considerable information has accumulated with this approach. In determining function, the physical properties of lipids are as important as their chemical properties. Yet how the physical properties of lipids measured *in vitro* relate to their *in vivo* function- there is little understanding.

In addition, the physical properties of lipids have been ignored in many *in vitro* studies. This approach has considerable limitations when applied to lipids because genetic approaches are generally the most useful in studying *in vivo* function. First, in order to make mutants with altered lipid composition and genes that do not encode lipids and the genes encoding enzymes along a biosynthetic pathway must be targeted.

Therefore, the results of genetic mutation are indirect, and many times far removed from the primary one. Second, providing the permeability barrier of the cell is a primary function of most membrane lipids. Therefore, before other functions of a particular lipid are uncovered alterations in lipid composition may compromise cell permeability. A lipid is essential for cell viability one may learn from genetics but for other requirements never learn the molecular bases. To establish the biosynthetic pathways of most of the common lipids over the past 20 years, genetic approaches have been successfully used.

To manipulate the lipid composition of cells without severely compromising cell viability: the challenge is to use this genetic information. The combination of the genetic approach to uncover phenotypes of cells with altered lipid composition and the dissection *in vitro* of the molecular basis for the phenotype has proven to be a powerful approach to defining lipid function, in those cases where this has been possible.

The more difficult the application of the genetic approach, the more complex the organelle content and accompanying membrane structure of a cell. Therefore, from genetic manipulation of prokaryotic and eukaryotic microorganisms, the most useful information to date has come. However, in more complex mammalian systems, these basic molecular principles underlying lipid function will be generally applicable.

8.4.1. Alteration of Lipid Composition in Bacteria

Eugene Kennedy and coworkers subsequently verified using genetic approaches the pathways for the formation of the major phospholipids (PE, PG, and CL) of *E. coli* . In defining new roles for lipids in cell function, the design of strains in which lipid composition can be genetically altered in a systematic manner has been very important.

Due to the barrier function of the outer membrane, unlike many other mutations affecting the metabolic pathways in *E. coli*, mutants in phospholipid biosynthesis cannot be bypassed by supplementation of the growth media with phospholipids. Therefore, the isolation and study of *E. coli* phospholipid auxotrophs have not been possible. Even under laboratory conditions, mutants in all steps of phospholipid biosynthesis were thought to be lethal, apart from the synthesis of CL. For cells unable to synthesize CD diacylglycerol to date, no growth conditions have been established.

A suppressor of this mutation has been identified null mutants in the PGS A gene (encodes phosphatidyl glycerophosphate synthase) that cannot synthesize PG and CL are lethal. The major outer membrane lipoprotein precursor accumulates in the inner membrane and apparently kills the cell, which depends on PG for its lipid modification, in such mutants.

This indicates that PG and CL are not absolutely required for viability, only for optimal growth cells unable to make this lipoprotein are viable but are temperature sensitive. However, for the proper membrane association and function of peripheral membrane proteins, the anionic nature of these lipids (apparently substituted by increased levels of PA) is necessary.

Based on initial mutants carrying temperature-sensitive alleles of the genes (*pss*A and *psd*) encoding their respective biosynthetic enzymes the amine-containing lipids, PS and PE, were also thought to be essential. However, by adding Ca^{2+}, Mg^{2+}, and Sr_{2+} in millimolar concentrations to the growth medium the growth phenotype of these mutants (as well as pssA null strains) with reduced amine-containing lipids could be suppressed.

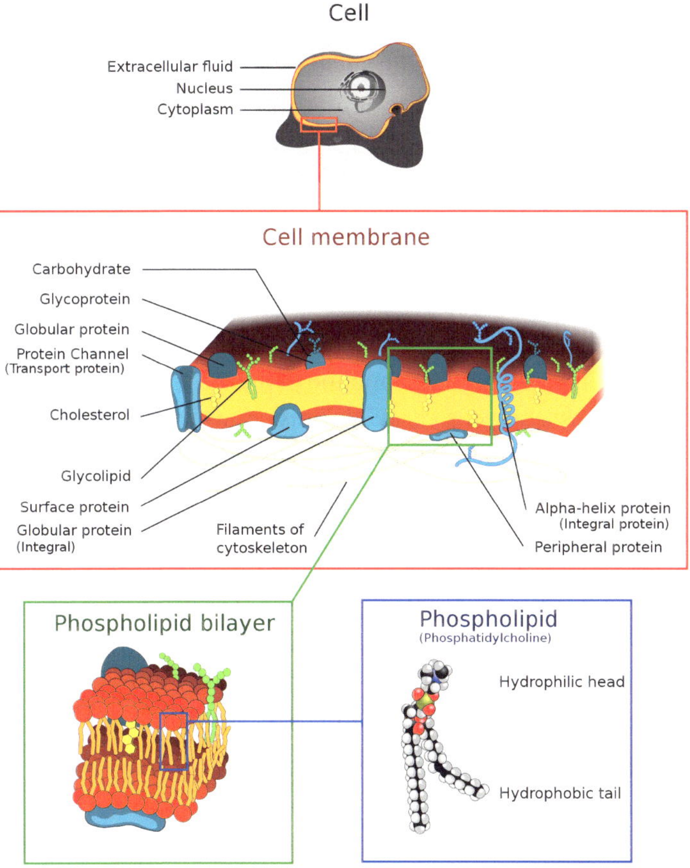

Figure 8.4. The cell membrane.

Source: Image by Wikimedia commons

These mutants, although viable, have a complex mixture of defects in in sugar and amino acid transport, reduced growth rate, loss of outer membrane barrier function, cell division, defects in energy metabolism and dis-assembly of membrane proteins. The key to defining new functions for the anionic and zwitterionic phospholipids of *E. coli* was the design of strains in a systematic manner in viable cells in which the content of PG/CL and PE could be regulated.

Isopropyl-β-thiogalactoside levels in the growth media was placed under the control of the exogenously regulated -controlled promoter *lacOP* (promoter of the lac operon) and controlled by the pgsA gene (encoding the phosphatidyl glycerophosphate synthase). In a dose-response manner variation in PG plus CL levels were correlated with the functioning of specific cellular processes both in vivo and in vitro to determine lipid function.

Similarly, by comparing phenotypes of cells with and without PE or by placing the *pss*A gene (encoding PS synthase) under exogenous regulation, the involvement of PE in function was uncovered. Therefore, to define potential lipid involvement in cellular processes *in vivo* such genetically altered strains have been used as reagents.

8.4.2. Alteration of Lipid Composition in Yeast

The genetics of lipid metabolism and the pathways of phospholipid synthesis and in yeast *Saccharomyces cerevisiae* are as well understood as in *E. coli*. Like those in the *E. coli*, yeasts have pathways for PE and PG synthesis. Rather than from one PG to another PG as in bacteria CL synthesis in all eukaryotes involves the transfer of a phosphatidyl moiety from CD diacylglycerol to PG.

In addition, for the synthesis of PI, PE, and PC including the methylation of PE to form PC, yeasts utilize the mammalian pathways. For viability, all gene products necessary for the synthesis of diacylglycerol, CDP-diacylglycerol, and PI are essential in yeast.

If the growth medium is supplemented with ethanolamine in order to make PE and PC, PS synthesis is not essential. Since pssl (encodes PS synthase) mutants also lack a sphingolipid degradative enzyme, however, PE is required, that generates ethanolamine internally, require ethanolamine supplementation.

By no gene products involved in lipid metabolism are encoded the mtDNA, which in *Saccharomyces cerevisiae* encodes eight proteins (subunits I, II, and III of cytochrome c oxidase, cytochrome b, the 3 subunits that make up the F0 component of ATP synthase, and the VAR1 gene product which is part of the mitochondrial ribosome).

They are encoded in nuclear genes and imported into the mitochondria the enzymes necessary for synthesis of PE from PS, and for PG and CL.

Mitochondrial function is not required for null mutants of CRDL (encodes CL synthase) to grow normally on glucose. However, they grow slower on non-fermentable carbon sources such as glycerol or lactate.

Therefore, for optimal mitochondrial function, CL appears to be required but for viability, but it is not essential. However, due to a null mutation in the PGS1 gene (encodes phosphatidyl glycerophosphate synthase) lack of PG and CL synthesis results in the inability to utilize nonfermentable carbon sources for growth.

To support the import of all nuclear-encoded proteins the mitochondrial membrane potential is reduced to near undetectable levels although remains sufficient. Similar effects are seen in mammalian cells with a mutation in the homologous PGS1 gene. The surprising consequence of the lack of PG and CL in yeast is the lack of cytochrome c oxidase subunit IV, which is nuclear-encoded by translation of mRNAs as well as of four mitochondria-encoded proteins (cytochrome b and cytochrome c oxidase subunits I-III).

These results indicate that either lack of these lipids indirectly affects both mitochondrial and cytoplasmic mRNA translation or some aspects of translation of a subset of mitochondrial proteins (those associated with electron transport complexes in the inner membrane but not ATP metabolism) require PG and/or CL.

8.5. ROLE OF LIPIDS IN CELL FUNCTION

There are at least two ways by which lipids can affect protein structure and function and thereby cell function. Protein function is influenced by specific protein-lipid interactions that are dependent on the chemical and structural anatomy of lipids (chelating properties, backbone, headgroup, alkyl chain length, ionization, degree of unsaturation and chirality).

However, protein function is also influenced by the unique self-association properties of lipids that result from the collective properties (thickness, shape, fluidity, packing properties) of the lipids organized into membrane structures.

8.5.1. The Bilayer as a Supramolecular Lipid Matrix

The L_α state of the membrane bilayer is essential for cell viability, as established by biophysical studies on membrane lipids coupled with biochemical and genetic manipulation of membrane lipid composition.

However, membranes are made up of a vast array of lipids that can assume individually different physical arrangements and have different physical properties to contribute collectively to the final physical properties of the membrane.

Animal cell membranes are exposed to a rather constant temperature, pressure, and solvent environment and do not change their lipid make up dramatically. In the L_α phase over the variation in conditions, we encounter the complex membrane lipid composition including cholesterol that stabilizes mammalian cell membranes.

Microorganisms are exposed to a broad range of environmental conditions so have developed systems in order to exist in the L_α phase for changing membrane lipid composition. Yet all biological membranes contain significant amounts of non-bilayer-forming lipids.

8.5.2. Selectivity of Protein-Lipid Interactions

A specific phospholipid requirement has been determined for optimal reconstitution of function *in vitro* for more than 50 membrane proteins. The number is in the hundreds if one considers specific lipid requirements for membrane association and activation of peripheral membrane proteins.

In a very complex environment, integral membrane proteins fold and exist and have three modes of interaction with their environment. To the water milieu, where they interact with water, solutes, ions and water-soluble proteins where the extramembrane parts are exposed.

To the hydrophobic-aqueous interface region, part of the protein is exposed. Within the approximately 30 Å thick hydrophobic interior of the membrane, the remainder of the protein is buried. Peripheral membrane proteins may spend part of their time completely, or even partially inserted into the membrane and are recruited to the membrane surface, in response to various signals. New roles for lipids have uncovered by genetic approaches coupled with *in vitro* verification of function.

Most exciting has been results from X-ray crystallographic analysis of membrane proteins which have revealed lipids in specific and tight association with proteins. An α-helix of 20-25 amino acids which is sufficient to span the 30 Å core of the bilayer is the predominant structural motif for the membrane-spanning domain of membrane proteins. A β-barrel motif is also found to a lesser extent.

8.5.3. Translocation of Proteins across Membranes

Movement of proteins across membranes is required to transfer a protein from its site of synthesis to its site of function. The transfer of hydrophobic and hydrophilic segments of proteins through the hydrophobic core of the membrane is included in this process. Considerable attention has recently been focused on the role of lipids in this process.

The *in vivo* evidence for the participation of anionic phospholipids in protein translocation was obtained from experiments with *E. coli* mutant strains defective in the biosynthesis of PG and CL. In these cells the *in vivo* translocation of the outer membrane precursor proteins, prePhoE and proOmpA, was severely hampered.

To fine-tune the level of PG in the membrane, the expression of the *pgs*A gene (encodes phosphatidyl glycerophosphate synthase) placed under control of the lac promoter/operator and the amount of PG was directly proportional to the translocation rate of the proteins.

As a translocation ATP-driven motor, SecA, which is a peripheral membrane protein moves secreted proteins through the membrane translocation pore composed of SecY and two other membrane proteins, SecE and SecG. SecY for high affinity binding to the membrane, for membrane penetration, and for high level ATPase-dependent function. SecA requires both anionic phospholipids and pore components.

Functional reconstitution of purified and mutated SecYEG complex from *E. coli* and *Bacillus subtilis* into liposomes of defined lipid composition revealed an absolute requirement for PG. Translocation activity was proportional to the amount of PG in reconstituted proteoliposomes and optimum activity was obtained only with the specific lipid composition of each organism.

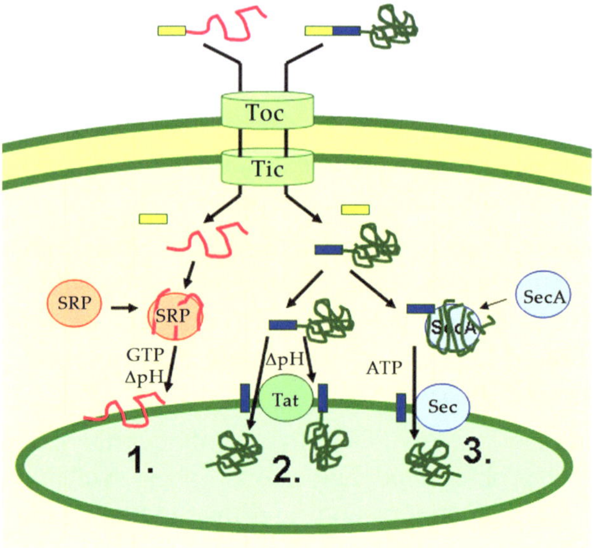

Figure 8.5. Protein targeting the thylakoid diagram.

Source: Image by Flickr

The N-terminal signal peptides of *E. coli* secreted proteins possess at least one positively charged amino acid. The membrane protein translocation efficiency is dependent on both the number of positive charges and the anionic phospholipid. Photocrosslinking of the secreted proteins with the fatty acid chains of membrane lipids demonstrated direct contact between the signal peptide and lipids during early stages of protein translocation. For protein translocation of proteins across the membrane of *E. coli*, non-bilayer-forming lipids are also required.

In the presence of Mg^{2+} or Ca^{2+} ,the only non-bilayer-forming lipid in *E. coli* mutants lacking PE is CL. From PE-deficient cells is reduced with divalent cation depletion but can be enhanced by inclusion of Mg^{2+} or Ca^{2+} protein translocation into inverted membrane vesicles prepared.

By incorporation of non-bilayer PE (18 : 1 acyl chains), translocation in the absence of divalent cations is restored but not by bilayer-prone PE (14 : 0 acyl chains). These results indicate that lipids with a tendency to form non-bilayer structures, to provide a necessary environment for the translocase of proteins across the membrane.

8.5.4. Assembly of Integral Membrane Proteins

In the insertion and organization of integral membrane proteins much less is known about the role of phospholipids. With several c~-helical transmembrane domains spanning the membrane bilayer, most such proteins are organized. These helices are connected by extramembrane loops alternately exposed on either side of the membrane.

In bacteria, these proteins generally obey the 'positive inside' rule in which those extramembrane loops on the cytoplasmic side of the membrane have a net positive charge and the loops exposed to the exterior are either neutral or negatively charged. These positive loops are anchored to the inside of the cell by interaction with anionic phospholipids indicated by the variation in the positive charge density of these loops as well as the anionic phospholipid content of the membrane.

8.5.5. Lipid-Assisted Folding Of Membrane Proteins

The membrane clearly serves as the solvent within which integral membrane proteins fold and function. However, to guide and determine final membrane protein structure and organization do lipids act in more specific ways? In the folding of membrane proteins recent evidence supports a role for lipids analogous to that of protein molecular chaperones.

By interacting with non-native folding intermediates, molecular chaperones maintain the native conformation are required to facilitate the folding of substrate proteins but do not interact with native or totally unfolded proteins.

In assisting the folding of specific membrane proteins lipids that fulfill these requirements by simply providing a solvent for the folding process have been termed 'lipo-chaperones' to distinguish their function.

The major evidence for the existence of lipo-chaperones comes from studies on the requirement for PE in the assembly and function of lactose permease (LacY) of *E. coli*. By alternating cytoplasmic and periplasmic loops, LacY is a polytopic membrane protein with 12 connected transmembrane-spanning domains.

LacY carries out transport of lactose (active transport) in an energy-independent mode to equilibrate lactose across the membrane (facilitated transport) or by coupling uphill movement of lactose against a concentration gradient with the downhill movement of a proton coupled to the proton electrochemical gradient across the membrane.

A lipo-chaperone role for PE in the assembly of LacY came about from studies. For studying LacY, many biochemical and molecular genetic resources are available including antibodies and engineered derivatives of LacY.

To study the requirement for PE in the assembly of LacY *in vivo* and in isolated membranes the availability of viable *E. coli* strains that either lack PE, or in which the level of PE can be regulated , the development of a blotting technique termed an 'Eastern-Western' made possible the screening for lipids affecting the refolding of LacY *in vitro* or the conformation of LacY made *in vivo*.

Lipids are first applied to a solid support such as nitrocellulose in the Eastern-Western procedure. Next, the protein of interest is transferred to the lipid patch proteins subjected to sodium dodecyl sulfate polyacrylamide gel electrophoresis are transferred by standard Western blotting techniques to the solid support.

As transfer continues, the sodium dodecyl sulfate is removed leaving behind the protein. Using conformation-sensitive antibodies or protein function by direct assay attachment of the refolded protein to solid support allows one to probe protein structure.

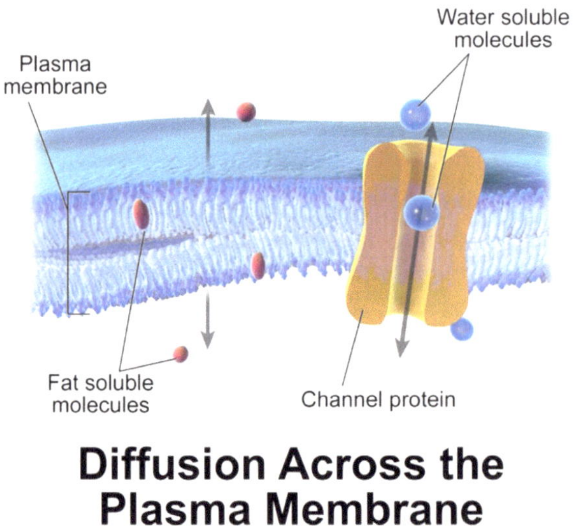

Figure 8.6. Diffusion across the plasma membrane.

Source: Image by Wikimedia commons

This combined blotting technique allows the detection of membrane protein conformational changes as influenced by individual lipids during refolding. Refolding from sodium dodecyl sulfate detects intermediate and late steps of folding since many membrane proteins including LacY retain significant amounts of secondary structure even in sodium dodecyl sulfate.

The initial observation that PE was required for LacY function was concluded from studies of reconstitution of transport function in sealed vesicles made of purified LacY and lipid. Both active and facilitated transport was restored when reconstituted in lipid mixtures containing PE. In mixtures containing only PG and/or CL or even PC, only facilitated transport occurred. Using mutants lacking PE the physiological importance of PE for LacY function was established. Cells lacking PE only displayed facilitated transport even though bioenergetic parameters of the membrane were normal while LacY expressed in PE-containing cells had full transport function.

It was established that LacY assembled in the presence, but not in the absence, of PE display's 'native' structure using Western and Eastern-Western blotting techniques and a conformation sensitive antibody.

Even when PE is completely removed, LacY maintains its native structure, and in the absence of PE LacY is originally assembled which is restored to its native structure by partial denaturation in sodium dodecyl sulfate followed by renaturation in the presence of specifically PE (or other primary amine-containing phospholipids such as PS). LacY assembled (either *in vivo* or *in vitro*) in membranes lacking PE is restored to the native structure by post-assembly addition of PE to the membranes. Furthermore, by simple exposure to PE LacY is extensively denatured in urea-sodium dodecyl sulfate (which eliminates most secondary structure) and cannot be renatured.

Hence, PE assists in the folding of LacY by a transient noncovalent interaction with a late-folding and non-native intermediate thereby fulfilling the minimum requirements of a molecular chaperone. .

In the absence of PE, the molecular basis for the loss of native structure and function of LacY assembled is a topological mis-assembly of the protein. The first six transmembrane domains and the loops that connect them assume an inverted topology with respect to the plane of the membrane bilayer in the absence of PE. Periplasmic loops become cytoplasmic and cytoplasmic loops become periplasmic. There is a correction of topology and regain of full transport function, if LacY is first assembled in the absence of PE and then post-assembly PE synthesis is initiated.

These results dramatically illustrate the specific effects of membrane lipid composition on both structure and function of membrane proteins,. For regulatory roles of lipids in cell processes, the ability of changes in lipid composition to affect such large changes in protein structure has important implications.

For example, they encounter different membrane lipid compositions that might affect protein structure in dramatic ways, as eukaryotic proteins move through the secretory pathway, to turn on or turn off the function. Local changes in lipid composition may also result in similar changes in structure and function.

8.5.6. Lipid Domains

An important role in cell function is played by compartmentalization of many biological processes such as biosynthesis, degradation, energy production, and metabolic signaling. Multiple membrane structures, cytosol versus membrane are all utilized to compartmentalize functions in subcellular organelles. Due to lipid polymorphism and differences in steric packing of the acyl chains, lipid mixtures made up of defined lipids undergo phase separations. Supporting the existence of segregated domains within the bilayer mixtures of bilayer and non-bilayer lipids undergo multiple phase transitions.

If their hydrophobic domains are the same, in model systems amphipathic polar lipid analogs self-associate into domains even if their polar domains carry the same net charge. By orderly packing of the hydrophobic domains, therefore, headgroup repulsive forces can be overcome. Between phospholipids and sphingolipids there is considerable acyl chain mismatch, i.e., as compared to the longer (20-24 for the acyl group) saturated chains of sphingolipids, phospholipids tend to have shorter acyl chains (16-18) with higher degrees of unsaturation.

Near the physiological temperature of 37°C, naturally occurring sphingolipids undergo the $L\beta$ to $L\alpha$ transition while this transition for naturally occurring phospholipids is near or below 0°C. Therefore, the more laterally compact hydrophobic domains of sphingolipids can readily segregate from the more disordered and expanded domains of unsaturated acyl chains of phospholipids.

By specific polar headgroup interactions, lipid segregation can also be facilitated, particularly intermolecular hydrogen bonding to other lipids and to protein networks involving hydroxyls, phosphates, amines, carbohydrates, and alcohols. Its high pK_2, as noted earlier, maybe the basis for the the hydrogen bonding properties of CL and the formation of clusters of CL in natural and artificial membranes.

8.5.7. Cytokinesis

The function of cytokinesis is to divide one cell into two by building a membrane barrier between the two daughter cells. To form a cleavage furrow in eukaryotic cells, the interaction of actin filaments with myosin filaments applies tension to the membrane, which gradually deepens until it encounters the narrow remains of the mitotic spindle between the two nuclei.

Phospholipids play an essential role in the division processes in eukaryotic cells. Lipid bilayer phospholipids in biological membranes are distributed asymmetrically between the inner and outer leaflets of the plasma membrane of eukaryotic cells. To the inner leaflet and PC and sphingomyelin are enriched and in the outer leaflet in the plasma membrane of eukaryotic cells PE and PS are localized.

It was demonstrated that PE is exposed on the cell surface of the cleavage furrow of eukaryotic cells at the final stage of cytokinesis using a cyclic peptide highly specific for binding to PE. The formation of a long cytoplasmic bridge between the daughter cells results from the immobilization of cell surface PE by the PE binding peptide inhibited disassembly of the contractile ring. With the disappearance of exposed PE, removal of the peptide from the surface of arrested cells allowed cell division to proceed. Furthermore, as a variant was isolated that was resistant to the cytotoxicity of the PE-binding peptide, a mutant cell line defective in PE biosynthesis.

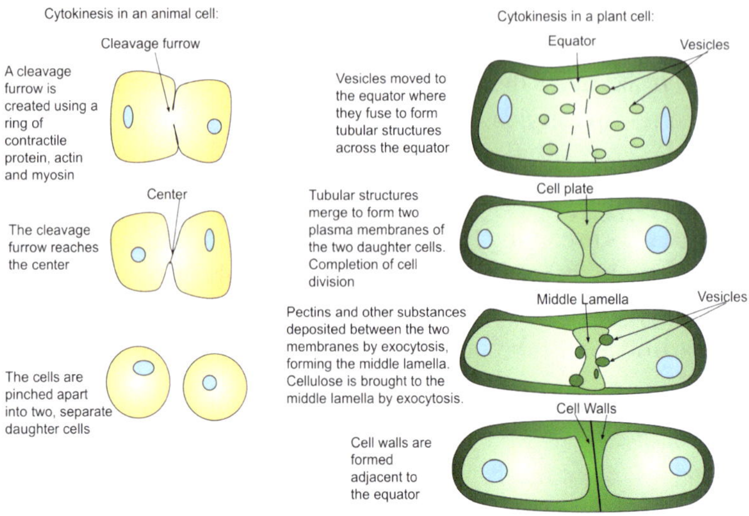

Figure 8.7. The comparison of the process of cytokinesis in plant and animal cells.

Source: Image by Wikimedia commons

For normal growth and cell division, this cell line required either PE or ethanolamine. Between the two daughter cells, in ethanolamine-deficient medium these mutant cells were arrested with a cytoplasmic bridge. The addition of PE or ethanolamine restored normal cytokinesis.

These findings provide the evidence that trans bilayer movement of PE at the cleavage furrow contributes to the regulation of cytokinesis. By organization of FtsZ protein monomers midway between the poles of the cell division is initiated after genome duplication, in *E. coli*. A series of proteins to the division site that brings about constriction and eventually cell division is recruited by this protein ring.

An *E. coli* mutant completely lacking PE propagates as long filamentous cells. The FtsZ ring fails to constrict when the FtsZ protein complex is recruited to the division site.This phenotype is not observed in strains with specific defects in other steps of phospholipid biosynthesis.

Prokaryotic membranes also appear to have an asymmetric enrichment of PE on the inner leaflet of the cytoplasmic membrane, although not firmly documented. For such cytoskeletal organization in the completion of cytokinesis in prokaryotic cells as well as mammalian cells, it is likely that PE is essential.

8.6. LIPID METABOLISM IN PLANTS

Most animals, including humans, depend on these lipids as a major source of calories and essential fatty acids because plants produce the majority of the world's lipids. As signal molecules, and as a form of stored carbon and energy, like other eukaryotes, plants require lipids for membrane biogenesis.

In addition, soft tissues and bark each have distinctive protective lipids that help prevent desiccation and infection. A profound effect on both gross lipid composition and the flow of lipid within the cell is contained by the presence of chloroplasts and related organelles in plants.

As in animals and fungi, fatty acid synthesis occurs not in the cytosol, but in the chloroplast and other plastids. To multiple compartments, and the complex interactions between competing pathways acyl groups must then be distributed and are a major focus of plant lipid biochemists. Rather than phospholipids, it is also significant that the lipid bilayers of chloroplasts are largely composed of galactolipids. As a result, galactolipids are the predominant acyl lipids in green tissues and probably on earth. On the world economy and human nutrition, plant lipids also have a substantial impact.

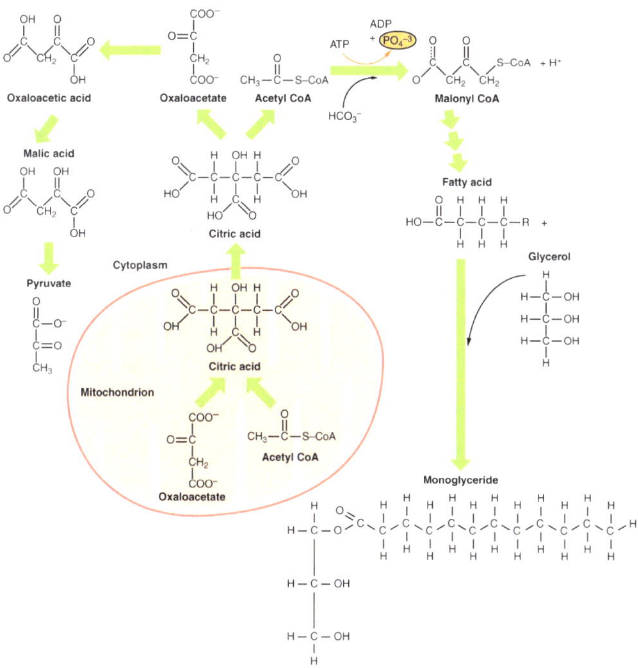

Figure 8.8. Lipid metabolism.

Source: Image by Wikimedia commons

More than three-quarters of the edible and industrial oils marketed annually are derived from seed and fruit triacylglycerols. Plants store more carbon as carbohydrates than as lipids on a whole organism basis. Plant requirements for storage lipid as an efficient, light-weight energy reserve are less acute than those of animals, since plants are not mobile, and since photosynthesis provides fixed carbon on a regular basis.

Finally, hundreds of genes for plant lipid biosynthesis, utilization and turnover required have now been cloned. To design new, more valuable plant oils in addition to providing valuable information on enzyme structure and function, these genes are being exploited. As DNA microarray and other genomic technologies develop, the coordination of lipid metabolic genes with each other and with their potential regulators may also become better understood.

8.7. PLANT LIPID GEOGRAPHY

8.7.1. Plastids

The ultrastructure of a plant cell differs from that of the typical mammalian cell in three major ways although all eukaryotic cells have much in common. By the cellulosic cell wall, the plasma membrane of plant cells is shielded, preventing lysis in the naturally hypotonic environment but making the preparation of cell fractions more difficult.

The nucleus, cytosol and organelles are pressed against the cell wall by the tonoplast,the membrane of the large, central vacuole that can occupy 80% or more of the cell's volume. Finally, all living plant cells contain one or more types of plastids. A circular chromosome presents in multiple copies, the plastids are a family of organelles containing the same genetic material.

Young or undifferentiated cells contain tiny pro-plastids that, depending on the tissue, may differentiate into photosynthetic chloroplasts, carotenoid-rich chromoplasts, or any of several varieties of colorless leucoplasts, including plastids specialized for starch storage.

These different types of plastids have varying amounts of the internal membrane but invariably are bounded by two membranes which may be interconverted *in vivo*. The internal structure of chloroplasts is dominated by the flattened green membrane sacks known as thylakoids. The thylakoid membranes are the site of the light reactions of photosynthesis and contain chlorophyll.

8.7.2. Endoplasmic Reticulum and Lipid Bodies

The endoplasmic reticulum has traditionally been viewed as the primary source of phospholipids in plant cells. All the common phospholipids can be produced by microsomal fractions except for cardiolipin.

Although plastids formed on acyl lipid substrates and are not typically exported, and they do have the ability to synthesize polyunsaturated fatty acids. Thus, for developing seeds that store large quantities of 18: 2 and 18:3 the endoplasmic reticulum desaturation pathways are of particular importance.

In microsomes, pathways to produce unusual fatty acids found primarily in seed oils have likewise been described. Not surprisingly, the endoplasmic reticulum also appears to be instrumental in the formation of the triacylglycerols themselves and the lipid bodies in which they are stored.

8.7.3. Mitochondria

The plant mitochondrion is probably the most thoroughly organelle investigated with respect to lipid metabolism next to plastids and the endoplasmic reticulum. Its ability to synthesize phosphatidylglycerol and cardiolipin is well established.

From the plastids or the ER, although most fatty acids for mitochondrial membranes are imported recently mitochondria have been shown to synthesize low levels of fatty acids from malonate. Octanoate serves as a precursor for the lipoic acid cofactor needed and is a major product of this pathway by glycine decarboxylase and pyruvate dehydrogenase.

8.8. FUTURE DIRECTIONS OF LIPIDS

In nature, the roles lipids play in cellular processes is as diverse as the chemical structures of lipids found. A diversity of lipid structure and physical properties is necessary to fill the broad range of roles that lipids play in cells although a sealed bilayer vesicle in solution can be formed by a single phospholipid can form.

With their often-harsh environments, lipid structures vary greatly from the archaebacteria, to the eubacteria that also must carry out a diversity of processes in one or two membrane structures, and to eukaryotic cells that have specialized organelles with different lipid compositions tailored to their function.

Because of the diversity of chemical and physical properties of lipids and the fact that each lipid type potentially is involved at various levels of cellular function, defining lipid function is a challenging undertaking. Biological membranes provide the means to compartmentalize functions, but at the same time, they perform many other duties and are flexible self-sealing boundaries that form the permeability barrier for cells and organelles.

Their physical properties directly affect these processes in ways that are often difficult to assess as a support for both integral and peripheral membrane processes. Each specialized membrane has a unique structure, composition and function.

Within each membrane exist subdomains as lipid rafts, lipid domains, and organizations of membrane associated complexes with their own unique composition. For integral and peripheral membrane proteins, as the organization site and many times are transient responding to cellular signals that can themselves be lipids. Lipids are integral components of stable complexes and serve specific structural roles by affecting protein conformation, by serving as the 'glue' that holds complexes together, or by providing the flexible interface between protein subunits. Lipids can also act in a more specific manner as molecular chaperones directing the attainment of final membrane protein organization and they provide the complex hydrophobic-hydrophilic solvent within which membrane proteins fold and function.

These diverse functions of lipids are made possible by a family of low molecular weight molecules that are physically fluid and deformable to enable interaction in a flexible and specific manner with macromolecules. They can organize into the very stable but highly dynamic supramolecular structures we know as membranes, at the same time.

Determining the function at the molecular level for the many lipid species already discovered will be the challenge for the future. Coupling genetic and biochemical approaches has been historically a very powerful approach to defining structure-function relationships of physiological importance.

This has proven to be very fruitful using this approach in microorganisms. As the sophistication of mammalian cells and whole animal genetics evolves, genetic manipulation coupled with biochemical characterization will begin to yield new and useful information on the function of lipids in more complex organisms.

The interest in understanding biodiversity that must be characterized structurally and functionally through the detailed characterization of the vast number of microorganisms will yield additional novel lipids. Finally, the role lipids play in disease will become more evident, as we discover more about the role of lipids in normal cell function.

8.9. CONCLUSION

The chapter explained the biochemistry of lipids. It also discussed the diversity in the structure of lipids. In this chapter, several properties of lipids in a solution have also been discussed. It also explains the engineering of the membrane lipid composition and the role of lipids in cell function. Towards the end of the chapter, it discussed the lipid metabolism in plants and plant lipid geography.

REFERENCES

1. Brown, H. and Marnett, L., 2011. Introduction to Lipid Biochemistry, Metabolism, and Signaling. *Chemical Reviews*, [online] 111(10), pp.5817-5820. Available at: <https://pubs.acs.org/doi/10.1021/cr200363s> [Accessed 3 July 2022].
2. D.E. Vance and J.E. Vance, 2002. *Biochemistry of Lipids, Lipoproteins and Membranes*. 4th ed. [ebook] Elsevier Science. Available at: <http://library.lol/main/98F4728DF0B7A4961056CBC00BE92D63> [Accessed 3 July 2022].
3. Calandra, S. and Tarugi, P., 1990. Biochemistry of lipids and lipoproteins. *Current Opinion in Gastroenterology*, [online] 6(3), pp.398-406. Available at: <https://journals.lww.com/co-gastroenterology/Citation/1990/06000/Biochemistry_of_lipids_and_lipoproteins.9.aspx> [Accessed 4 July 2022].
4. Carman, G., 1997. Biochemistry of lipids, lipoproteins and membranes edited by D. E. Vance and J. Vance, Elsevier, 1996. US$210.50 (hbk) (xxii + 553 pages) ISBN 0 444 82359 X. *Trends in Biochemical Sciences*, [online] 22(9), p.369. Available at: <https://linkinghub.elsevier.com/retrieve/pii/S0968000497827594> [Accessed 4 July 2022].
5. Higgins, J., 1992. Biochemistry of lipids, lipoproteins and membranes. *FEBS Letters*, [online] 310(1), pp.95-96. Available at: <https://febs.onlinelibrary.wiley.com/doi/abs/10.1016/0014-5793%2892%2981166-J> [Accessed 4 July 2022].

INDEX

K

L

M

N

O

P

Q

R

S

T

U

V

W

X

Printed in the United States
by Baker & Taylor Publisher Services